UNCERTAINTY AND THE PHILOSOPHY OF CLIMATE CHANGE

When it comes to climate change, the greatest difficulty we face is that we do not know the likely degree of change or its cost, which means that environmental policy decisions have to be made under uncertainty. This book offers an accessible philosophical treatment of the broad range of ethical and policy challenges posed by climate change uncertainty.

Drawing on both the philosophy of science and ethics, Martin Bunzl shows how tackling climate change revolves around weighing up our interests now against those of future generations, which requires that we examine our assumptions about the value of present costs versus future benefits. In an engaging, conversational style, Bunzl looks at questions such as our responsibility towards nonhuman life, the interests of the developing and developed worlds, and how the circumstances of poverty shape the perception of risk, ultimately developing and defending a view of humanity and its place in the world that makes sense of our duty to nature without treating it as a rights bearer.

This book will be of interest to students and scholars of environmental studies, philosophy, politics and sociology, as well as policy makers.

Martin Bunzl is Professor of Philosophy at Rutgers University, USA, where he founded the Rutgers Initiative on Climate and Social Policy.

Martin Bunzl brings philosophy out of the ivory tower and into our everyday lives. What responsibility do I have to my future self, 20 years from now? What responsibility do we, as a society, have to future generations and the natural environment? Bunzl compellingly makes the case that an understanding of philosophic issues is central to successfully addressing the most important question of our lives: How to live well in a world where our actions can inflict (or at least not alleviate) hardship and suffering on others?

Ken Caldeira, Carnegie Institution for Science, USA

Martin Bunzl takes the idea of climate change as a risk management problem seriously. In a book that is personal, as well as philosophically, politically, and scientifically well-informed, he asks us to see the challenge of climate change in collective terms. Our future is dark, according to the author, unless we can overcome our individualism and parochial concerns.

Dale Jamieson, New York University, USA

UNCERTAINTY AND THE PHILOSOPHY OF CLIMATE CHANGE

Martin Bunzl

Routledge
Taylor & Francis Group

LONDON AND NEW YORK

earthscan
from Routledge

First published 2015
by Routledge
2 Park Square, Milton Park, Abingdon, Oxon OX14 4RN

and by Routledge
711 Third Avenue, New York, NY 10017

Routledge is an imprint of the Taylor & Francis Group, an informa business

© 2015 Martin Bunzl

British Library Cataloguing-in-Publication Data

A catalogue record for this book is available from the British Library

Library of Congress Cataloging-in-Publication Data

Bunzl, Martin, author.
Uncertainty and the philosophy of climate change / Martin Bunzl.
    pages cm
  1. Climatic changes – Philosophy.   2. Environmental policy –
Forecasting.   I. Title.
QC903.B86   2015
363.738′7401—dc23   2014011784

ISBN: 978-1-138-79391-0 (hbk)
ISBN: 978-1-138-79392-7 (pbk)
ISBN: 978-1-315-76073-5 (ebk)

Typeset in Bembo
by Apex CoVantage, LLC

For Jonah

CONTENTS

LIST OF FIGURES AND TABLES

Figures

Tables

PREFACE

This idea for this book of essays was conceived during a period of greater optimism about the prospects for limiting climate change than that in which we find ourselves today. Because of the long gestation period for the manuscript, the focus of the argument reflects this shift. It is in the end a decidedly pessimistic argument about the prospects for collective action. That said, all is not lost. But as I argue in the penultimate chapter of the book, avoiding the risk of climate change will likely involve a massive effort to put things right after the fact. Doing that will involve extraordinary cost and even then we won't be able to put things back to the way they were. Nonetheless we will have a chance to avoid the potential for catastrophic outcomes. Collective action now would be a much cheaper and more effective way to do the same things. How and why such action is so hard to achieve is my concern in what follows.

I am very grateful for the release time from teaching my university and department gave me to think about climate change, and for the home Ruth Mandel of the Eagleton Institute of Politics gave me to do so while establishing the Initiative on Climate and Social Policy there.

My work was partly supported by the National Science Foundation (ATM-0730452) as well as Rutgers University.

The book would not have been possible without the help of an extraordinary number of people:

Alan Robock, who welcomed me into his climate research group and tutored me in the basics of geoengineering and volcanology.

Sally Baho, Huiyuan Feng, Ben Kravitz, Yuriy Levin, Charles Lin, Jianqiang Liu, Joe Reilly, and Clinton Smith, who each provided invaluable research assistance.

Shenyu Belski, Ying Chen, Feng Gao, Kejun Jiang, Yan Li, Dong Liu, Jiahua Pan, Yi Qi, Zhen Sun, Ailun Yan, Fuqiang Yang, Jie Yu, Tao Wang, Changhua Wu, Haiban

Zang, and Yisheng Zheng in Beijing; Reid Dechton, Lisa Friedman, Melanie Hart, Trevor Hauser, Andrew Light, Aldan Meyer, Jake Schmidt, and Harlan Watson in Washington; Neal Carter, Mike Childs, Nick Eyre, Michael Grubb, John Gummer, James Murray, Michael Pollit, Steve Rayner, William Rickett, and Bryony Worthington in the United Kingdom; and Adrienne Alvord, Ken Alex, Susan Frank, Susan Kennedy, Lawrence Lingbloom, Kip Lipper, Steve Magiglio, Jan Mazurek, Cliff Rechtshaffen, Nancy Skinner, and Terry Tamminen in California, who all provided me with essential insight into climate policy and negotiations that form the basis for Chapters Eight and Nine in this volume.

Audiences provided valuable questions at talks drawn from this material presented at: The AAAS, The AGU, Berkeley National Lab, IRADe (New Dehli), The ISS (Potsdam), Lawrence Livermore Labs, NYU, Rutgers, Princeton, TERI (New Delhi), and UCSD.

Thanks to Ben Kravis, Christopher Preston, Rupert Read, and Toby Svoboda for their careful read-through and helpful comments.

Some of the material in Chapters Seven, Ten, and Eleven is drawn from: Bunzl, Martin 2009. "The Tragedy of the Commons: A Reassessment," *Climatic Change*, 97: 59–65, Bunzl, Martin 2011. "Geoengineering Harms and Compensation," *Stanford Journal of Law, Science & Policy*, 4: 70–76, Bunzl, Martin 2009. "Geoengineering Research: Shouldn't or Couldn't?," *Environmental Research Letters*, 4: 1–3, and Bunzl, Martin 2007. "The Next Best Thing" in Anthony Appiah and Martin Bunzl (eds.), *Buying Freedom: The Ethics and Economics of Contemporary Slave Redemption*, Princeton: Princeton University Press, 235–248, and is used here with permission of the publishers.

A note on terminology. Much of what follows draws on the work of the Intergovernmental Panel on Climate Change (IPCC) which produces reports on climate change every seven years. Over the last three reports, the IPCC has used a series of economic scenarios and associated energy use to develop alternative models of anthropogenic greenhouse gas emissions into the atmosphere and the expected radioactive forcing they may be expected to produce. While in successive reports (AR3, AR4, and AR5) its projections have narrowed but not changed significantly, the names of the scenarios have changed over time. Thus the business as usual scenario was first termed A1C (because it assumed high coal use), became A1F1 (broadening to focus reliance on all fossil fuels), and most recently RCP8.5 (based on its associated rate of radiative forcing per square meter). In what follows, to keep things simple, I introduce uniform terminology for the three scenarios I make most use of. For those familiar with IPCC nomenclature the (rough) translation schema is as follows:

BAU (Business as Usual) = A1C, A1F1, RCP8.5
BOT (Betting on Technology) = A1T, (roughly) RCP 6.0
BTN (Back to Nature) = B1, RCP4.5

based on the radiative forcing associated with each scenario as shown in Figure P.1.

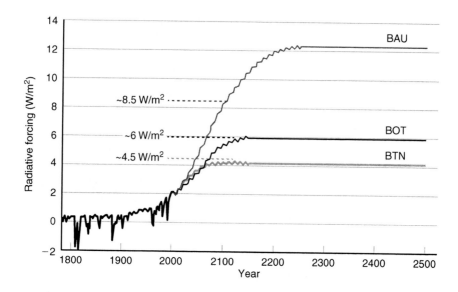

FIGURE P.1   Radiative Forcing (IPPC 2013, Chapter 1, 147, Box 1.1 Figure 1).

Finally a note on evidence. What follows makes use of three different kinds of argument, each of which have their own standards of adequacy. First, there are thought experiments which are only used to clarify conceptual arguments. These have no evidential requirements. Second, there are scenarios of unknown but non-zero probability that would be cataclysmic if they came about, and hence, I argue, deserve to be taken seriously as long as they are merely physically possible. Finally, there are arguments that depend on the plausibility of underlying empirical claims and their associated probabilities.

Reference

IPCC 2013. *Climate Change 2013: The Physical Science Basis*, http://www.climatechange2013.org/images/report/WG1AR5_Chapter01_FINAL.pdf, accessed February 13th, 2014.

INTRODUCTION

I

There was a time when I thought it would be a mistake to have children. In the late 1970s, the probability of large scale nuclear contamination of the world, sometime during the likely lifespan of the next generation, seemed high enough to make being born a dubious proposition. What favor would it be to bring a child into world with the prospect of a nuclear winter on the horizon?

And then everything changed. Suddenly, with detente and then the end of the USSR, the prospect of wholesale nuclear interchange evaporated. Yes, there was, and still is, a risk of local interchanges. Those would still have devastating effects for worldwide agricultural production for two decades (Robock and Toon 2012) but they would not end all life. And so I had children. But my sense of (relative) equanimity has been short lived. My nuclear fears for my offspring have now been replaced by climate fears. Not for them, their children or even their grandchildren, but for the generations that will follow.

But is there any reason to think that climate change could have catastrophic consequences akin to a nuclear winter? James Hansen (2008) thinks there is. Hansen et al.'s (2005) modeling suggests the existence of amplifying feedbacks that can realize runaway scenarios in which the planet stabilizes in a (cold) state akin to Mars at $-50\,°C$ or a (hot) state akin to Venus at $+450\,°C$. In the Mars scenario, increasing surface albedo as the planet cools is the amplifying force. In the Venus scenario, increasing water vapor as the planet warms drives things. But there is a big difference between these two scenarios. Runaway cooling has happened before (most recently 640 million years ago) but is subject to change as weathering and rising atmospheric CO_2 reverse the process. However Hansen argues that there is no such reversing process in the case of runaway heating, hence the potential for mass permanent extinction.

The crisis of the nuclear and the crisis of climate have something in common, beyond their overwhelming destructive potential. They both prompt a sense of utter helplessness for us as individuals, a sense of something totally out of our control. More than that, it is not just that there is a sense that they are out of our personal control, but that in the end that they are just not controllable, a sense that collective rational action is somehow beyond our reach.

If anything, the dissolution of the nuclear crisis underscores this sense. After all, it did not come about in any planful way. There was no grand bargain. Instead, the world stumbled into a good outcome while dodging many bullets along the way. There was a sense that a bad outcome was much more likely than a good one. But if a bad outcome was more likely that a good one in the nuclear case, what is the balance between the good and the bad outcomes in the climate case? When the United States and the USSR faced each other with massive nuclear arsenals, the strategy of mutually assured destruction made for an all or none calculus by design. There would be peace or the apocalypse. But climate change is a more complicated matter. If Hansen's Venus scenario is a possibility, it is only one of many possibilities, even if it is the worst of them. Our difficulty is that we don't know the likely degree of climate change nor its cost for all of these possibilities taken together. Policy questions about avoiding climate change thus involve decisions under uncertainty that are much more complex than those involving nuclear war.

II

One way to think about such decisions is to do our best to fix the value of both the likelihood of climate change and its cost. Such a project is far from straightforward because it not only touches on our epistemic limitations but also on assumptions about the valuation of present costs versus future benefits. In Chapter One I begin by contrasting making such decisions about avoiding climate change with the way we think about decisions involving risk and benefits closer to home. An obvious difference would seem to be this. My decision not to smoke involves a calculus of self-interest. The risks and benefits are mine. Avoiding climate change is a calculus between our interests now and those of others in future generations. While that may not make a difference when it comes to collective self-interest, I argue that it does when it comes to individual risk perception, which is crucial for support of government action for the collective good. Collective self-interest or not, when it comes to climate change, how are we to evaluate its risks and costs? In Chapter Two I examine an approach that seems to sidestep this challenge. Never mind what the chances of anthropogenic causes of climate change really are, or the likely costs. However small the chances are, if it is possible that climate change could produce catastrophic outcomes, à la Venus scenario, should we not avoid them, whatever the cost? On one variant of this argument, the cost of such an outcome would be infinite if we consider all of the future generations who would be deprived of existence, and so paying any price would seem "rational" as long as it is less than the cost of extinction. But what if the cost is not infinite, but merely very very high? (For after all, life on earth will end eventually, if only because the sun becomes a red giant.) Then the calculus becomes much more complicated. How can I be sure the price to avoid extinction is not more than the cost of extinction? I defend the view that even though the costs of extinction are not infinite, they are high enough that we can effectively ignore this question. But to do so, I surely need to argue that we have an obligation to future generations of human beings.

The standard literature on future generations assumes there will be future beings but that their identities and numbers are a function of our actions today. But what if my actions now cause total human extinction in the future? Then whom have I wronged? Constructing a philosophical framework to support the intuition that here too I have done wrong, even though I wronged no one, is the challenge I take on in Chapter Three. The attempt to do so lays the groundwork for a much more controversial project, to make sense of the idea that our actions can also wrong nonhumans.

Whatever our obligations to others may be, if we have any, what then? In Chapters Four and Five I examine the question of how we should divide up those obligations. In contemporary political discourse the answer is quite simple, the problem of climate was caused by the Developed World and so it is up to the Developed World to solve the problem. Moreover, the common perception is that since clean energy is available, it is just a question of whether we are willing to pay the higher costs for using it and who should pay those costs. Whoever does pay, the interests of the Developing World and the Developed World are taken to be aligned in that we all lose in the face of climate change. As such, avoiding climate change trumps development for everyone, including the poor. And since the poor are overwhelmingly concentrated in the Developing World, so too, in the end, avoiding climate change will trump development in the Developing World.

But scratch the surface and none of this is as straightforward as it may seem at first blush. I argue that who did what, when, and who owes whom what is not only complicated morally but very dependent on where we stand in time. A few years from now, things will look very different because of the growing output of the Developing World. So from where we look at things in time matters. The idea that clean energy can be had, but that it is only a matter of cost, not only makes assumptions about the availability of such energy but the rate at which the infrastructure to distribute it can be deployed in comparison to the rate of growth of energy demand. The idea that short-term interests may, in fact, trump long-term interests for the poor on the assumption that a clean energy supply cannot reconcile these two interests occupies much of Chapter Four. There I examine the problem looking from the outside in, even when it comes to judgments of rational choice. What happens if we look at the same choices from the inside out?

In Chapter Five I examine how the circumstances of poverty shape the perceived calculus of risk, and in doing so, further tip the balance in favor of the short term for better or worse.

But the poor are not alone in tipping this balance. As I argue in Chapter Six, another way this happens is the result of the how we approach the distribution of risk collectively through insurance and disaster relief programs, and the way in which that affects the rationality of our decision making individually.

What if we look instead to our political leaders to save us from our individual, narrow, short-term interests? In Chapter Seven I examine the standard argument that, absent a comprehensive agreement, the problem of the Tragedy of the Commons seems to cast doubt on how likely this is to happen. That is the idea that some countries' self-constraint will be exploited by others so there is no motivation for any country to show constraint, leading to disaster for all. But I argue that this is the wrong way to analyze the reluctance of countries to negotiate about climate, and indeed that the widespread belief that climate change is an instance of the Tragedy of the Commons limits our horizon of possibility.

In theory at least, some countries (notably China and the United States) are large enough in their share of the world economy that they could unilaterally implement policies that would force others to follow. In Chapter Eight I examine why it is that China and the United States have failed to do so and will likely continue to fail to do so if it means slowing economic growth. On the other hand, some economies are too small to implement policies on their own that force others to follow. Yet some of them (California and the United Kingdom) have nonetheless chosen to implement such policies. In Chapter Nine I examine their reasons for doing so and ask whether theirs is a model that could be generalized to the rest of the world. At the risk of depriving the reader of a sense of suspense, the answer turns out to be that they likely could not be so generalized.

Avoiding the risks of climate change through political leadership will only happened if it is achieved collectively. But that will only be possible if political leaders take a multigenerational view of interests and don't set the discount rate on the future too high. Absent this I argue we will only be moved to act collectively when the accumulated effects of climate change are clearly attributable to it and widely distributed enough in time and place to affect most of the globe. But by then, it may be too late to act without avoiding serious effects of climate change. What then, if anything, is to be done? In Chapter Ten I argue that this logic should drive us to take the need for air capture geoengineering seriously, despite its costs and potential risks.

Whether we act now by reducing carbon output or act later by attempting to remove it from the atmosphere presumes we are willing to act. The upshot of my argument is that we have a moral imperative to act, if not for our own species then for others. That may be the end of it when it comes to a philosophical argument. But in the last chapter I examine the interplay between ethics and psychology to examine why following such an imperative seems so hard.

III

In David Guggenheim's 2006 documentary, *An Inconvenient Truth* (for print version see Gore 2006), Al Gore shows a picture of the land in Tennessee on which he grew up:

> You look at that river gently flowing by. You notice the leaves rustling with the wind. You hear the birds; you hear the tree frogs. In the distance you hear a cow. You feel the grass. The mud gives a little bit on the river bank. It's quiet; it's peaceful.

Gore speaks movingly about his desire not to see the land's beauty ravaged by climate change. But in doing so, he allows the impression that all we need to do is to make a few changes and the rest will be business as usual. For us, and him on his land. Change my light bulbs. Drive less. Heat my house less. Fly less. The Sierra Club gives me a list of ten things. I am advised to plant a tree in my garden. Others have longer to do lists for me. Puffing out its chest, *Vanity Fair* (Porter 2006) demands another forty things. (I should forgo preheating the oven.) Not to be outdone, the *Palm Beach Post* (Schwed 2007) offers ninety-nine prescriptions! (Use a hand potato masher instead of an electric one.) George Marshall (2007, 135) says I ought not to think of any of this as a sacrifice. He says I will feel proud. My new life style "will be a statement of who I am – a smart aware person living in the 21st century."

Al Gore and me, standing shoulder to shoulder. Why am I so unmoved? I want to be moved. I want to move. Yet here I sit. Unmoved. My lethargy might be because I really don't think my actions will make much of a difference. But I don't think my voting makes a difference. Yet the same thought does not stop me voting. Of course I only have to vote once in a while, so it is an act of minimal inconvenience. Is it that what I am asked to do here is so inconvenient and complicated?

If that is not bad enough, others demand even more of me. We need to change our whole outlook on life to save the planet. Gus Speth (2008) says I have to stop looking at nature as a means to my ends. I am too materialistic and too individualistic. Bill McKibben (2006) says I have to reintegrate human society and nature and foreswear anthropocentrism for a "biocentric" world view. I am told I should embrace a humbler world. If I listen to Speth and McKibben, I need to turn my life upside down. Even if I wanted to do that, I don't even know how to begin. The contours of my life are sown into a web of relations that makes such a change hard to contemplate except as a fantasy. I give everything away, sever all ties, live in a shack, tend my fields, and collect firewood. Even if that is fine for some, it is not for me.

Al Gore whispers in my ear: *"Ignore McKibben and Speth! They are naysayers and luddites. Walden Pond romantics! Stick with me. Together we can solve this problem. Yes, big changes are needed, but that does not mean our way of life has to change. All we need to do as a nation is* . . . to commit to producing 100 percent of our electricity from renewable energy and truly clean carbon-free sources within 10 years. . . . When President

John F. Kennedy challenged our nation to land a man on the moon and bring him back safely in 10 years, many people doubted we could accomplish that goal. But 8 years and 2 months later, Neil Armstrong and Buzz Aldrin walked on the surface of the moon. . . . We must now lift our nation to reach another goal that will change history. Our entire civilization depends upon us now embarking on a new journey of exploration and discovery. Our success depends on our willingness as a people to undertake this journey and to complete it within 10 years. Once again, we have an opportunity to take a giant leap for humankind." (This Gore quote and those that follow are partly fictional [in italics] and partly based on his speech (Gore 2008), given on July 17, 2008 at Constitution Hall, Washington.)

Is it that simple? Merely a matter of will and our (American) ingenuity? Gore makes it seem almost un-American to wonder if there really is a technical solution merely waiting for the ambitious to grab. For him there is *always* a technical solution to every problem. That is what makes America America! And if my friends still die of cancer decades after a Kennedy-like moon program was declared to defeat it, we just have not tried hard enough. But if I am allowed to stamp my foot and command discovery or innovation, I too can solve any problem. And in the long run, no doubt we can solve the problem. But as Keynes reminded us, in the long run we are all dead.

Gore points his finger at me. "*Maybe you didn't listen to my speech carefully enough. I said we can solve this problem in 10 years. All we need to decide to do it do it!*"

I don't get it, don't facts intrude? Where do we store the power for use at night when there is no wind or light? What about China and India's rising energy needs? Gore casts a condescending eye on me. "Of course there are those who will tell us this can't be done. Some of the voices we hear are the defenders of the status quo – the ones with a vested interest in perpetuating the current system, no matter how high a price the rest of us will have to pay. But even those who reap the profits of the carbon age have to recognize the inevitability of its demise. As one OPEC oil minister observed, 'The Stone Age didn't end because of a shortage of stones.'"

Right. It ended because a more productive cost effective technology came along. Al Gore in his bully pulpit, stamping his foot can't change the fact that that is just what we lack for now and the foreseeable future. "*You know, if you had paid attention you would have heard me call for CO_2 caps and revenue neutral taxes! It is all so simple.*"

I affect a professorial mien to add some gravitas to my brief. "China will move 240 million people from the country to the cities by 2025 and two thirds of its population will live in urban areas. City people in China use more than twice the energy of country people. Over 400 million people in India population lack electricity. India's national goal is to be 100% electrified by 2030 and its electrical demand is projected to grow five to six fold by 2050" (Woetzel et al. 2009, Remme et al. 2011).

"*Look*," says Gore, "it is also essential that the United States rejoin the global community and lead efforts to secure . . . a global partnership that recognizes the necessity of addressing the threats of extreme poverty and disease as part of the world's agenda for solving the climate crisis."

IV

It all seems too easy. In India 68.7% of the population lives on less than $2 a day (World Bank 2012). It just signed a 25 year contract to import 9 billion tons of coal annually from the United States (*Courier Journal* 2012).

Six hundred million of us got the life we have because of our industrial revolution that would not have been possible without energy to fuel it. Between 1820 and 2004, United States primary energy consumption grew from .837 quadrillion BTU to 100 quadrillion BTU, even as energy consumption per real dollar of GDP fell over time: from 66,690 to 9,400 BTU per dollar of GDP. By comparison, China's 2006 energy consumption was 13,799 BTU per real dollar of GDP (PPP). (See U.S. Energy Information Administration 2010a, 2010b, 2010c, and Maddison 2007, 379.) Now 6 billion more want to improve their lives. Even if they don't reach our standard of living, the numbers alone will drive the growth in energy demand, which is expected to increase by 45% in the next 20 years of which 70% will come from Developing World (see Figures I.1 and I.2).

At the same time, renewable and nuclear energy cannot be expected to grow from current levels (7% of current total energy) to fill this demand. Looking just at electricity, the anticipated mix can be seen in Figure I.3.

When it comes to economic growth, of course, rising demand for electricity is only a portion of the source of the problem of CO_2 emissions caused by fossil fuels, as seen in Figure I.4.

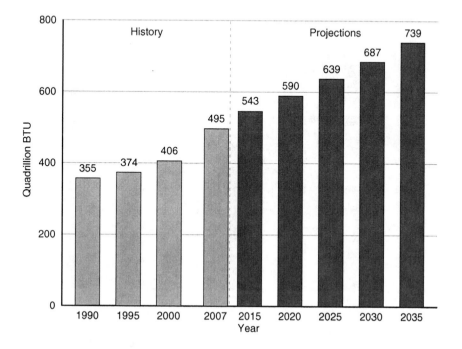

FIGURE I.1 World Energy Consumption 1990–2035 (U.S. Energy Information Administration 2010d)

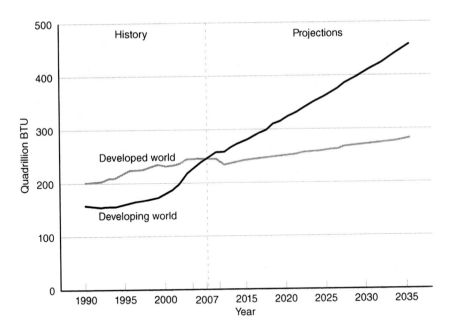

FIGURE I.2 Developing and Developed World Energy Consumption 1990–2035 (U.S. Energy Information Administration 2010d)

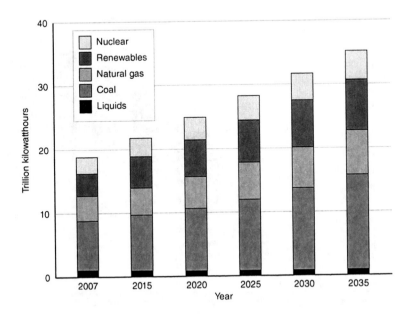

FIGURE I.3 World Electrical Generation 2007–2035 (U.S. Energy Information Administration 2010d)

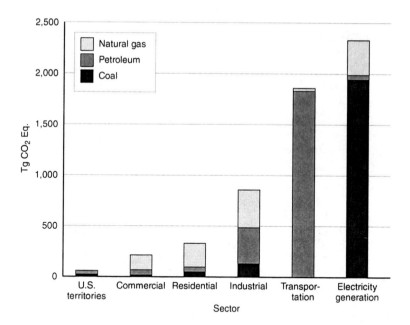

FIGURE I.4 Emissions of CO_2 by Sector and Fuel Type (EPA nda)

Note: Electricity generation also includes emissions of less than 0.5 Tg CO_2 Tq. from geothermal-based electricity generation.

Nor is fossil fuel the only source of the problem, even if it is the major source (87%) of it. Demand for land use, iron and steel production, cement production, lime manufacture, ammonia production, solid ash manufacture, and ammonia production all rise with economic growth and make up the remaining 13%.

But if that is not bad enough, there is an even worse scenario. If household size is caused by rising household income, we cannot expect population to stabilize (at roughly 9 billion people) without such growth (see for example Murdoch 1980 and Galor 2011).

As such, either way we face increasing demand for energy, for *ongoing* population growth itself implies *ongoing* growth in energy demand. The only way to limit rising energy demand *in the long run* is to stabilize population. But if raising household income is a condition on stabilizing population and the only way to raise household income requires increased energy use, then, short of a drastic reduction of population from current levels by disease or draconian social policies, the challenge of climate change is not simply whether or not we are willing to live simpler lives to avoid it. It is whether or not we can provide the needed energy for population levels to stabilize without producing damaging changes in the climate. As we will see, this is no trivial calculus when it comes to both considerations of both economics and of ethics. But, above all, it is a non-trivial calculus because of our limited knowledge about the risks of climate change itself.

That said, the Intergovernmental Panel on Climate Change (IPCC 2013 Technical Summary 50, 63) AR5 Synthesis Report's projections has a best estimate of a 3.7 °C rise in global mean temperature by the end of the 21st century, but a likely range between 2.6 and 4.8 °C under the "business as usual" scenario of rapid economic growth in which fossil fuels play a primary role (BAU), while on the same scenario, global sea level rise is projected to be between .45 and .81 meters with a best estimate of .62 meters. (See Table I.1.)

Beyond the end of this century the consequences of business as usual becomes even more stark with a projected range between 3–12.6 °C over 1986–2005 average global mean surface temperature with a best estimate of 8 °C (IPCC 2013, Technical Summary, 60). (See Figure I.5.)

TABLE I.1 Projected Change in Global Mean Surface Air Temperature and Sea Level Rise (IPCC 2013, Technical Summary, 90 based on Table TS.1)

Variable	Scenario	2046–2065		2081–2100	
		Mean	Likely Range	Mean	Likely Range
Global Mean	BTN	1.4	0.9 to 2.0	1.8	1.1 to 2.6
Surface	BIT	1.3	0.8 to 1.8	2.2	1.4 to 3.1
Temp. Change in °C	BAU	2.0	1.4 to 2.6	3.7	2.6 to 4.8
Global	BTN	0.26	0.19 to 0.33	0.47	0.32 to 0.62
Mean Sea	BOT	0.25	0.18 to 0.33	0.47	0.33 to 0.62
Level Rise in Meters	BAU	0.29	0.22 to 0.37	0.62	0.45 to 0.81

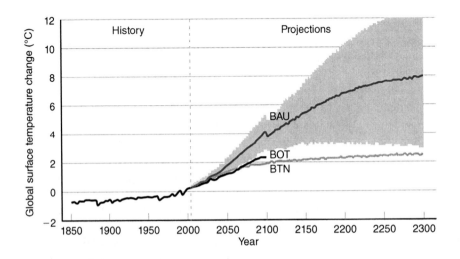

FIGURE I.5 Projected Global Annual Mean Surface Temperature Change by Scenario (IPCC 2013, 89, Table TS.1).

V

Naomi Oreskes and Erik Conway (2011) rightly claim that elimination of all doubt (in favor of certainty) is not part and parcel of the scientific process and go on to tell the sorry talk about the misuse of appeals doubt in scientific policy debates. Oreskes and Conway think that scientific knowledge is created by a consensus of scientific opinion, a view that correctly makes such knowledge still revisable in the light of new evidence. But the problem is that policy cannot always afford to wait for such consensus to develop. Indeed, when it comes to climate change, while there may be scientific consensus on the role of humans in causing temperature changes over the *last* 100 years, looking *forward*, our models become more uncertain the further out we project into the future in an attempt to predict how much temperature change there will be. In this sense, demanding scientific consensus is no better a require-ment than demanding the elimination of all doubt (unless it is consensus in favor of doubt itself). And whether the public accepts that there is scientific consensus about the causes of temperature changes in the past is irrelevant. What matters is on what basis we should decide on policy looking forward without much confidence about the severity of climate change in the future, not because of lack of consensus but because of the limitations of our current climate models. All of this is to say that doubt pervades our current climate science when it comes to confidence about the future. What stance should we take toward this uncertainty?

Now Oreskes and Conway would have us believe that we pay a high price for embracing doubt when it comes to rational decision making because "the outcome of a rational decision-theory analysis is that if your knowledge is uncertain, then your best option is generally to do nothing" (Oreskes and Conway 2011, 267). If that were correct then we ought to do nothing about climate risk since uncertain is just what our knowledge of the severity of climate change is. Oreskes and Conway would have us believe that "[i]f we didn't know that smoking was dangerous, but we did know that it gave us pleasure, we would surely decide to smoke, as millions of Americans did before the 1960s" (Oreskes and Conway 2011, 267). That is surely wrong. Despite the deliberate corporate attempts to fan doubts to undermine the evolving consensus about the risks of smoking that Oreskes and Conway detail in their book, we didn't need to *know* that smoking was dangerous to make a rational decision that it was not worth the risk that it *might* be dangerous and that those risks outweighed the benefits.

Waiting for scientific knowledge, for scientific consensus, before making policy choices is not a luxury we always have. Instead, we often need to make policy decisions when our knowledge is uncertain, but it does not follow that under such circumstances our "best option is generally to do nothing," as Oreskes and Conway suggest (Oreskes and Conway 2011, 267).

Your internist refers you to a specialist for a growth. The specialist recommends surgery and chemotherapy. You decide to get a second opinion. That specialist rec-ommends just surgery. It turns out there is no consensus among specialists. Or perhaps there is consensus about this: we just don't know whether surgery and che-motherapy produces better results than surgery alone. What should you do? How should you decide? Common sense would suggest it is a matter of how much risk and pain from the chemotherapy itself you are willing to bear in case the addition of it were to produce better results. You choice is about whether or not to pay a premium to (perhaps) increase your probability of a cure. In buying insurance we pay a premium to offset the costs of untoward outcomes should they occur. Here we pay a premium to reduce the chance of untoward consequences from happening in the first place. How to price the premium in either case is easy if we know both the risk and the cost of what we are trying to avoid. When we don't, the choice is still easy when the premium is trivial. (For example, my doctor has me take a low dose aspirin pill once a day because there is some evidence that it might reduce the risk of a heart attack.) But calculating the correct price to pay for avoiding climate change (the premium) is anything but trivial as our knowledge of the risks and costs of climate change are limited. How then to proceed?

References

Courier Journal 2012. "Kentucky Coal Producers Supply 9 Million Tons Annually India 25 Year 7 Billion Deal," www.courier-journal.com/article/20120815/BUSINESS/308150067/Kentucky-coal-producers-supply-9-million-tons-annually-India-25-year-7-billion-deal, accessed August 20th, 2012.

EPA nda. Climate Change Emissions, www.epa.gov/climatechange/emissions/CO2_human.html, accessed March 5th, 2012.

Galor, Oded 2011. "The Demographic Transition: Causes and Consequences," Working NBER Working Paper Series Paper 17057, www.nber.org/papers/w17057, accessed January 2nd, 2014.

Gore, Albert 2006. *An Inconvenient Truth*, Emmaus, PA: Rodale.

Gore, Albert 2008. *Al Gore's Speech on Renewable Energy*, www.npr.org/templates/story/story.php?storyId=92638501, accessed May 16th, 2011.

Hansen, James 2008. "Climate Threat to the Planet: Implications for Energy Policy and Intergenerational Justice," Lecture given Dec. 17 at the American Geophysical Union, San Francisco. (Slides posted at www.columbia.edu/~jeh1/presentations.shtml, accessed June 4th, 2011.)

Hansen, James, M. Sato, R. Ruedy, L. Nazarenko, A. Lacis, G. Schmidt, G. Russell, I. Aleinov, M. Bauer, S. Bauer, N. Bell, B. Cairns, V. Canuto, M. Chandler, Y. Cheng, A. Del Genio, G. Faluvegi, E. Fleming, A. Friend, T. Hall, C. Jackman, M. Kelley, N. Kiang, D. Koch, J. Lean, J. Lerner, K. Lo, S. Menon, R. Miller, P. Minnis, T. Novakov, V. Oinas, J. Perlwitz, D. Rind, A. Romanou, D. Shindell, P. Stone, S. Sun, N. Tausnev, D. Thresher, B. Wielicki, T. Wong, M. Yao and S. Zhang 2005. "Efficacy of Climate Forcings," *Journal of Geophysical Research: Atmospheres*, 110: D18104.

IPCC 2013. *Climate Change 2013: The Physical Science Basis*, www.ipcc.ch/report/ar5/wg1/#.UmGB0FCkojU, accessed October 18th 2013.

Maddison, Angus 2007. *Contours of the World Economy, 1–2030 AD. Essays in Macro-Economic History*, Oxford: Oxford University Press, table A.4.

Marshall, George 2007. *Carbon Detox*, London: Octopus.

McKibben, Bill 2006. *The End of Nature*, New York: Random House.

Murdoch, William 1980. *The Poverty of Nations*, Baltimore: Johns Hopkins Press.

Oreskes, Naomi and Erik Conway 2011. *Merchants of Doubt: How a Handful of Scientists Obscured the Truth on Issues from Tobacco Smoke to Global Warming*, New York: Bloomsbury.

Porter, Henry 2006. "Fifty Ways to Help Save the Planet," *Vanity Fair*, May, www.vanityfair.com/politics/features/2006/05/savetheplanet200605, accessed May 18th, 2011.

Remme, Uwe, Nathalie Trudeau, Dagmar Graczyk and Peter Taylor 2011. *Technology Development Prospects for the Indian Power Sector*, Paris: IEA.

Robock, Alan and Owen Brian Toon 2012. "Self-assured Destruction: The Climate Impacts of Nuclear War," *Bulletin of the Atomic Scientists*, 68 (5): 66–74.

Schwed, Mark 2007. "99 Things You Can Do to Save the Planet," *Palm Beach Post*, April 22, www.palmbeachpost.com/accent/content/accent/epaper/2007/04/22/a1d_earthday_lists_0422.html, accessed May 18th, 2011.

Speth, James 2008. *The Bridge at the End of the World*, New Haven: Yale University Press.

U.S. Energy Information Administration 2010a. *International Total Primary Energy Consumption and Energy Intensity*, www.eia.doe.gov/emeu/international/energyconsumption.html, accessed March 5th, 2012.

U.S. Energy Information Administration 2010b. *International Energy Statistics*, www.eia.doe.gov/cfapps/ipdbproject/iedindex3.cfm?tid=92&pid=46&aid=2&cid=r7,&syid=2004&eyid=2008&unit=BTUPUSDM, accessed March 5th, 2012.

U.S. Energy Information Administration 2010c. *Annual Energy Review 2010*, www.eia.gov/totalenergy/data/annual/pdf/aer.pdf, accessed March 5th, 2012.

U.S. Energy Information Administration 2010d. *International Energy Outlook 2010*, www.eia.doe.gov/oiaf/ieo/world.html, accessed May 16th, 2010.

Woetzel, Jonathan, Lenny Mendonca, Janamitra Devan, Stefano Negri, Yangmel Hu, Luke Jordan, Xiujun Li, Alexander Maasry, Geoff Tsen, Flora Yu 2009. *Preparing for China's Urban Billion*, McKinsey Global Institute, www.mckinsey.com/insights/urbanization/preparing_for_urban_billion_in_china, accessed July 1st, 2012.

World Bank 2012. *Poverty Headcount Ratio at $2 a Day (PPP) (% of Population)*, http://data.worldbank.org/indicator/SI.POV.2DAY, accessed August 20th, 2012.

1

RISK AND THE PERCEPTION OF RISK

I

Policy questions about avoiding climate change involve decisions under, at best, uncertainty. Our difficulty is that we don't know the likely degree of climate change, nor its cost. But these limitations on our knowledge may seem no different than those infecting many of the decisions we have to make on a daily basis.

I am 65 years old and have no long-term care insurance. Should I buy some? It is a question I have asked myself on and off for the last 10 years, but my decision remains unresolved. It is a hard decision to make, if only for psychological reasons. A desire for a quick death, or alternatively, immortality, clouds the decision. Bracketing psychological considerations, should I buy the insurance? As with all insurance, it is a decision under uncertainty. But it is one for which statistical data is available. There are enough data points that I can fine grain the data base as much as I want to derive a sample that is very much like me, even if it is not identical with me. Of course how much to fine grain is a matter of discretion, and one that may not matter too much depending on how little doing so changes the distribution of the results. Those results will give me a figure for the probability that I will need to use the policy at some point in future. What is much harder to decide on is this: what the value of the insurance is in dollars on the day I pay the premium. The benefit is in dollars in the future (if a payout is triggered), unless I have a stroke before the day on which the policy becomes effective. What discount rate should I apply to normalize the difference between the two? Suppose I pay $10,000 in a lump sum now and use the policy 10 years from now for a year, garnering $20,000 in benefits after which I die. The present value of that benefit is not $20,000 but the amount of capital that would yield $20,000 in 10 years if invested. How much that is depends on what I assume the rate of return to be. The present value (PV$) of $20,000 10 years from now is a function of the (inflation adjusted) interest rates (IR%):

IR%	PV$
4	13,500
5	12,000
6	11,000
7	10,000

Suppose I use the average rate of return on my retirement portfolio as a guide. Then the worth of the policy today is a function of the net present value of the payout for each year and the probability that I will use it in that year.

Are we then done? Not quite. There are additional considerations that may be relevant to how much to discount the value of a future payout. So far, all we have considered is essentially the inflation-adjusted opportunity cost of spending money on a premium instead of investing it. But the value of a dollar now as opposed to in the future is also a function of technology. Technological advances reduce the cost of living. So, $20,000 in 10 years may buy more than it does now, despite the effects of inflation. If I want to buy care worth $20,000 in 10 years based on current technology, on this consideration, I can plan to pay less for the same care in the future in real dollars. So I can further discount those dollars in the present. Finally, there is this to consider: maybe a dollar spent now is worth more to me than a dollar spent 10 years from now.

There is a good reason to ignore this last consideration, if only for my own (long-term) good. Money tends to burn a hole in my pocket and hence I tend to overdiscount the value of future money. But there other reasons that look more legitimate. I value life more now than I expect to value it in the future because I am younger and more vibrant now. I, literally, get more out of life. It is not that money spent caring for me in the future will be worth less than money spent on (say) wine and music now but that, *all other things being equal*, the value to me of a dollar then and now will be different. Still other reasons are not obviously illegitimate but not obviously legitimate either. I care about me now more than I care about the me that will be in the future. It is the kind of sentiment that tempts me not to go to the gym and diet. No doubt a life led now without excess will yield benefits to my future self, but I find it hard not to discount the value of my future self relative to my self now.

These complications notwithstanding, making such a decision under uncertainty for an individual is reasonably straightforward. And it might seem that extending the approach to collective decision making ought to be no less straightforward. That is certainly true if we stick to one generation. But as soon as the collective decision involves more than one generation things get more complicated.

Deciding whether or not to buy insurance involves two very distinct components: one is epistemic. That implicates both the matter of setting a value for the likelihood of particular outcomes and some features that go into setting a value on the discount rate, like the rate of return on capital and the rate of technological innovation. But the other component is purely valuational, as when I am forced to decide how much to care about the future as opposed to the present.

When we think of decision making across generations, the epistemic challenge does not change in kind but does in complexity. Time is our enemy here. The further our horizon of concern, the less confident we can be about our assignment of probabilities. It is one thing to extrapolate from a subset of the population now to another subset of the population now, and quite another thing to do the same thing across time and more so, the larger the swath of time. (For example, think how much both life expectancy and technology have changed in the last 100 years.) But when it comes to matters of valuation, a much deeper issue intrudes. My quandary about whether or not to buy long-term care insurance was a conflict between *my* interests now and *my* interests in the future. But when it comes to climate change, the tradeoff between interests is between us now and others in the future. That seems to turn a calculus of self-interest into a matter of altruism.

Still, to the extent that we think of our government as representing our collective self-interest, not only in time but across time, maybe the calculus of self-interest can be preserved. If we think of each generation as benefiting from the sacrifices of the previous generation, there is still always a temptation for each of us, as individuals, to break with the implicit compact between generations. We harvest the slow growing trees they planted, but fail to replant new ones. We benefit from the practical results of their costly theoretical research, but fail to fund such ongoing research at an appropriate level. We cross the bridges designed to last 100 years that they built, but fail to renew them as they become worn down. Yet knowing we are subject to such inclinations, we rely on our government to act on behalf of our better selves. We look to them to act in our self-interest where the "our" reaches out across generations. But be that as it may, such government action still has to depend on citizen support in the *here and now* and, whether based on self-interest or altruism, such support depends on our perception of how things are. So when a government acts in a way that may seem to be in the collective interest, it can't avoid dealing with how individuals see things.

II

I used to smoke and I enjoyed doing so very much. I gave up after my doctor told me it might kill me. "Might" was enough to convince me that the pleasure did not outweigh the risks. For all I know it may have turned out that his warnings were misplaced. Maybe the correlation between smoking and disease will turn out to be spurious – the result of a common cause. But the uncertainty gets swamped out by the costs of being wrong. The cost of death to me is high. High enough that even if there is only a small chance of my doctor being right, I want to avoid it. And the price of avoiding it, foregoing the pleasure, pales in comparison to that. Assume my actions are rational when it comes to smoking. What prevents their extension to averting climate change? One difference is this: the statistical data on the risks of smoking are well established in a way that is observer independent. They are a function of the frequency of death among smokers as compared to non-smokers.

Of course we have to control for all sorts of other things, but all other things being equal, "the effect of smoking on the chance of dying is similar to the effect of adding 5 to 10 years of age" (Woloshin et al. 2008, 845). Moreover, the Centers for Disease Control and Prevention estimates that on average, smoking reduces life expectancy by between 13 and 14 years (Centers for Disease Control and Prevention 2002). Yet when it comes to making an individual decision about whether to smoke or not, what matters is not what the observer independent risk is, but on what the *perceived* risk by the individual is. That subjective perception of risk (and with it cost) is what matters when it comes to making a decision based on the costs and benefits. So, even if what I will call the objective risk of not acting to avoid climate change might be the same as the objective risk of continuing to smoke, the perception of risk, the subjective risk, might be different. And indeed, the literature on risk perception gives ample reason to think that it will be different.

Beginning in 1978, Fischhoff et al. (1978) used factor analysis in an attempt to study how hazard is perceived. Their results support the view that much more is going on than a simple estimate of probabilistic frequencies. Instead, the ordering of risk for different activities is a function of two separate dimensions of perception, broadly: unfamiliarity and dread. The first provides an ordering in terms of risks that are perceived as:

> voluntary, immediate, known, controllable and old versus involuntary, delayed, unknown, uncontrollable, and new.

While the second is along a dimension between of risks perceived as:

> not certain to be fatal, common, and chronic versus certain to be fatal, dreadful, and catastrophic.

This provides for a distribution over hazards of the two-dimensional space shown in Figures 1.1a and b.

In this two dimensional space, the lower left quadrant represents the lowest area of perceived risk. Notice that smoking is located in that quadrant. Its relative position compared to other hazards (like satellite crashes) reflects how at variance subjective judgments of risk are as compared to an objective ordering.

That said, where we act to diminish risk, our risk-avoidance actions are not necessarily reflected in our ordering of risk perception. For example, people are more likely to stop smoking than flying, even if they (incorrectly) rate the latter as riskier than the former. But that should not be surprising, because risk avoidance is not just a function of risk perception; other considerations come into play as well. After all, the cost and inconvenience of foregoing air travel is high.

Is the difference between my attitude to risk avoidance in climate change as compared smoking that I assess the hazard as lower, or is it that here too other considerations (like the perceived cost and inconvenience of avoiding the risk of climate change) are in play? Of course, these are not mutually exclusive, but even if we bracket such considerations for now, a salient complication is that we need to

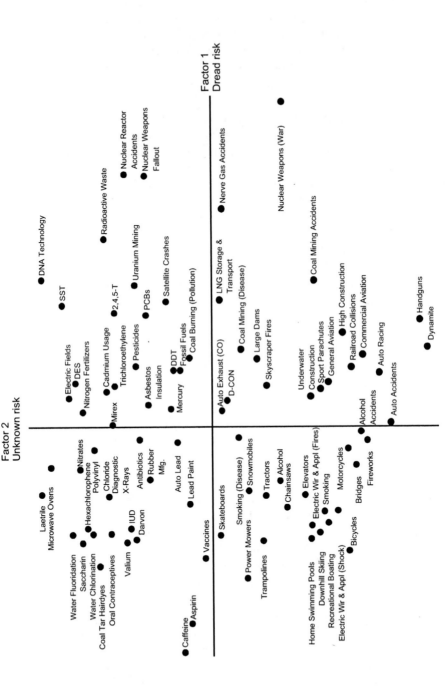

FIGURES 1.1a,b The Perception of Hazard (Slovic et al. 1980)

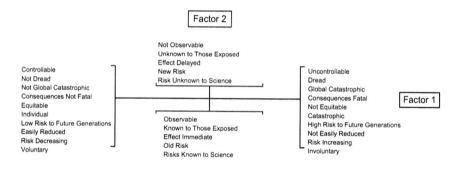

FIGURES 1.1a,b (*Continued*)

distinguish between consideration of risk avoidance of climate change as a question about me or as a collective question. The latter is surely the interesting of the two. But then we are comparing how important I think it is to act individually to avoid smoking as compared to how important I think it is to act collectively to avoid climate change. But if we focus on *risk perception*, the two are still comparable, for what we are interested in comparing is subjective risk perception. And that is still an individual affair whether the subject involves individual or collective action or choice.

If we apply the components of the two dimensional model of risk perception, where should we expect the risk of climate change to fall in that space? Swim et al. (2009) suggest that:

> To the extent that individuals conceive of climate change as a simple and gradual change from current to future values on variables such as average temperatures and precipitation, or the frequency or intensity of specific events such as freezes, hurricanes, or tornadoes, the risks posed by climate change would appear to be well-known and, at least in principle, controllable and therefore not dreaded.
>
> (Swim et al. 2009, 40–41)

But of course the risks of climate change are distributed over different outcomes, ranging from the relatively benign to the very bad. "Dread" is more likely when we focus on the very bad rather than the benign or even a weighted average of the two. Which do we focus on and which should we focus on? Smoking has many damaging effects on the body, but not all of them are fatal. For example, smoking postmenopausal women have lower bone density compared to non-smokers as well as an increased risk for hip fracture (U.S. Department of Health and Human Service 2001).

But although smoking may be bad for your health even if it does not kill you, in thinking about smoking, what we focus on are the catastrophic outcomes. And we do in other cases as well. Not wearing seat belts increases the risk of death but there are consequences less extreme than death; seat belts also lower the chances of injury. Yet while we focus on the former not the latter when it comes to seat belts,

TABLE 1.1 Willingness to Pay for Renewable Energy (Rabe et al. 2011)

	2008	2009
Nothing additional each year	22%	33%
1-49 dollars a year	16%	31%
50-99 dollars a year	17%	13%
100-249 dollars a year	13%	11%
250-499 dollars a year	10%	3%
500 dollars or more a year	7%	2%
Not sure	15%	6%

we don't seem to do so that when it comes to climate risk. A good way to gauge this is by assessing willingness to pay for policies that reduce carbon output. For example, in a poll for Brookings, Rabe and Borick (2011) asked respondents what they would be willing to pay annually for more renewable energy to be produced (see Table 1.1). That seems to reflect anything but an attitude of serious concern, even if we bracket those who are unwilling to pay anything at all as skeptical that climate change is real.

What is going on? On one dimension of risk, climate change ought to count as worse than smoking or not wearing seat belts. It is less voluntary, immediate, known, controllable, and old, rather than involuntary, delayed, unknown, uncontrollable, and new. Is the issue the second dimension of risk perception? Is it because it is treated as common and chronic rather than certain to be fatal, dreadful, and catastrophic? But if so why? This is just a restatement of the issue. Climate change has dreadful worst case outcomes even if they are not certain. Why do we not focus on them then in the way that we do with wearing seat belts whose worst case outcomes are also uncertain?

Perhaps the difference hinges on something beyond the two-dimensional model: namely, the role of experience of small probability events as opposed to analytic inferences about them. Experience amplifies risk judgments beyond their objective consequences (see for example, Slovic et al. 1986). An important difference between not wearing seat belts and climate change is that we have the opportunity to vicariously experience the consequences of the first but not the second. Think only of how we feel as we drive past the scene of a serious car crash, or even see a severely damaged car on a flatbed tow truck. (The same is true when we hear of or read about someone dying of lung cancer.) On the other hand, by and large, the (catastrophic) consequences of climate change are only available to us in abstract terms.

Abstraction enters into the picture in another way as well. The potential damage of climate change will not only fail to affect me directly, but it won't even affect my grandchildren with whose interests I identify, even if they don't exist yet and hence their identities are not fixed. I can imagine grandchildren, and in doing so my imagination is rich enough that I can even fantasize about their well-being and worry about that. My will contains a provision to provide for their college tuition.

If they are born, in all likelihood they will have children, my great grandchildren. What about *their* well-being? I have made no provision for them, and that is telling. For unlike my relationship to my prospective grandchildren, my relationship to my prospective great-grandchildren is purely abstract. If I was rich enough, perhaps abstraction would suffice in motivating me to setup a dynasty trust, but the key issue is how few generations it takes for an emotional bond to thin out into one that is purely abstract: abstract enough that caring itself becomes abstract and correspondingly thin as well for most of us.

Still abstract or not, the weakness of my concern for my far removed (hypothetical) descendants needs to be reconciled with the fact that "high risk to future generations" figures as one of the features of what produces higher scores along one of Fischhoff et al.'s (1978) dimensions of risk. But there is a difference: in the case of Fischhoff et al.'s model of risk assessment, high risk to future generations is assumed to be an assessment about a risk to us that also ramifies into future generations, in contrast to one that only is a risk to us. That is a difference between nuclear reactor accidents (above and beyond the numbers) and trampoline accidents. But the case of climate change does not fit that contrast. Extreme climate change is not a risk to us that ramifies to future generations but just a risk for future generations.

III

The temptation may be to fan the flames of risk perception in the here and now by insisting that climate change is producing change now that places us at risk. This was the strategy pursued most notably by Al Gore in *An Inconvenient Truth* (Gore 2006). It is a troubling temptation for two reasons. First, because of the inherent variability of climate, attributing any changes in the intensity or frequency of short-term weather patterns based on climate change is statistically problematic over short periods of time. The amount of "noise" means that you need a strong "signal" to have a chance of statistical significance, notwithstanding recent claims (see Hansen et al. 2012). Perhaps that concern can be appropriately parried in nonacademic settings by hedging the claims being made. But even with that in hand, to what end should we deploy such claims? Gore's aim was to induce fear, which certainly mirrors a popular approach when it comes to smoking in attempts to amplify the perception of risk. However, here too there is an important difference. In the case of smoking, fear is deployed to change behavior that is under the control of the individual. The individual has, as it were, a flight response available. But in the case of climate things are more complicated since collective action is needed to make a difference. Absent a collective response, what are the consequences of inducing fear? The danger here is that prolonged experience of fear, without a behavioral option to reduce it, can produce a psychological response that hurts rather than helps by being inhibitory (Moser 2007, Moser and Dilling 2004).

But if not fear, then what? Moser (2007, 71–75) argues for the imperative to make what she calls "supportive or enabling" conditions available concurrently with fear appeals, for the importance of "fostering hope" and "envisioning a future worth fighting for." Yet this construction assumes our identification with that future, and that is just what we lack, at least when it comes to the far off future, irrespective of the content, be it fear or hope. But it is only in the far off future that concerns about extreme climate change really engage.

In one sense there is a parallel to smoking: smoking was not a risk to me as a young person but to me as an old person. Indeed the challenge of programs to discourage smoking in young people is to see the risk to their future selves as a risk to themselves in the here and now. Swim et al. (2009, 44) suggest that we can learn from this parallel by arguing that successful programs to reduce smoking involve reducing the high degree of discounting for costs in the future and perhaps we can strive to do the same for the risks of climate change.

But in fact, the case of smoking teaches us a quite different lesson. Even if it may be rational not to smoke on long-term grounds, the convincing reasons not to smoke, for young people, may not be based on considerations of rationality per se. The reasons not to smoke may be our long-term health, but that does not mean a campaign to win such support needs to be based on those considerations. And as with smoking so too with climate.

Smoking and climate change are of a piece as compared with not wearing seat belts in the sense that the latter presents dangers in the here and now as well as the future, while smoking and climate change present dangers in the future alone. Still, there is a seemingly salient difference between smoking and climate change as well; the "me" that is danger of the effects of smoking is the "me" in the here and now. But in the case of climate change, the challenge is whether I will identify with the future generations that are at risk for harm. But is that a real difference? There is both a psychological basis and a philosophical basis for denying that there is.

Most 20 year olds have difficulty with acting on behalf of their future 80-year-old selves. But the idea that the difficulty is because of (over) discounting assumes the 20 year old identifies with the future 80 year old as "his" (or "her") future self. But what if the problem is of identity, not discounting? That may simply be a limitation of the imagination of a 20 year old but it might also express something more existentially substantial. Of course the 20 year old does not know what his 80-year-old self will be like, nor even if his 80 year old self will be alive. These are epistemic quandaries. They represent limits on what the 20 year old can know. But there is also an ontological quandary. The likes of the 80 year old, let alone whether he is alive, are not set 60 years ahead, unless we live in a much more deterministic world than I think we do. But even if we do live in such a world, there is a more interesting ontological feature of this as well that is philosophical in nature.

What makes the 20 year old and the 80 year old the same person?

In a psychological sense they may not be. Consider the 80 year old, looking back on his life, looking back on his 20-year-old self. In this direction the epistemology is settled, as is at least one feature of the ontology: what has happened has happened. There are no more indeterminacies in this direction. But the 80 year old may look back at the 20 year old and declare, "That was not me – I am now a very different person." Enough has changed so that he no longer feels any sense of communion between himself as he is now and his earlier self. But in what sense are they the same even if such a sense of communion obtains?

What should we say about this ontological notion of sameness? What makes this question especially complicated is that it implicates the problem of personal identity in the face of change. The 80 year old is both physically and mentally different from the 20 year old. In what sense are they the same person despite that change? Now one (radical) response to this question, counterintuitive as it may seem, is to simply deny the notion of personal identity itself. This view is largely due to Derek Parfit (1984) who champions the deflationary position that we don't need a concept of personal identity to make sense of our experience. Our experience is consistent with the notion that all that matters is psychological continuity and connectedness. What we are beyond that is not really important. There need be no identity relation between the me today and the me last year as long as we are connected by memory.

Now we normally think of the implications of Parfit's view as backward looking. The 80 year old is connected to the 20 year old not by a relation of identity but continuity. But in a way, the forward looking implications are more radical. The 20 year old, contemplating the 80 year old, is to take account of the interests of someone for whom he now (at 20 years old) will be a source of memories and connectedness. Now that sounds like a relationship not different in kind, but only in strength, from the 20 year old's relationship to future generations that are descended from him. In this sense, perhaps the challenges are the same. If that is right, before we worry about discounting, we should worry about the problem of connectedness looking forward for both us relative to our future selves and us relative to future generations. What if that is the situation of the 20 year old smoker? Parfit (1984, 319–320) argues that smoking as a young person is wrong because it poses risks for one's older self. But since we are not assuming any identity here, this argument is no different from one against smoking because it places others around us at risk. That poses an altruistic moral argument as the alternative to an argument based on self-interest.

That said, I am not totally indifferent between my future selves (even if they are not identical with me) and other future selves, or between future generations descended from me as opposed to others. My pension fund and life insurance policy have a beneficiary: my future self and not you, the reader (unless you happen to be my spouse). So, for better or worse, there *is* a psychological reality here that needs to be taken account of, even if Parfit is right and it rests on a philosophical confusion. In the end it is the psychology not the philosophy that matters if what we care about is how people make decisions now about risk that influences outcomes in the future. But what the philosophy can do is undermine the assumptions we make in trying to understand the psychology. What the philosophy does here is to undermine the ease with which we assume the straightforwardness of the notion of self-interest when it comes to the idea of rational self-interest.

If we look at antismoking campaigns, however, it turns out that long-term self-interest and altruistic moral argument are not in fact the only alternatives, especially when it comes to appeals to young people. The most obvious disincentive in the "here and now" is, of course, taxation, which has been demonstrated to be more effective than any kind of advertising campaign (Hu et al. 1995). But taxation is not the only alternative. Within advertising campaigns, there was a shift in the 1980s from an emphasis on long-term health effects to short-term effects like bad breath and sports performance. More recently, campaigns have attempted to capitalize on the desire of young people for independence by making them aware of attempts by tobacco companies to foster smoking dependence (Farrelly et al. 2003). Focus group research has suggested that these later campaigns are more effective than earlier short-term health campaigns (Goldman and Glantz 1998). Farrelly et al. (2005) reported a drop in smoking from 1999–2002 from 25.3% to 18% of which 22% was attributable to public campaigns.

The lesson of antismoking campaigns is that winning acceptance for public policy does not need to rely on the twin pillars of appealing to rational self-interest or altruism. But if we are prone to resist embracing this lesson it is because it goes against our conception of ourselves to think of our behavior and attitudes as being changed in ways that we are not a party to. Of these ways, the most prevalent revolve around the centrality of conformity. Even when conforming behavior is through the presentation of information directly and explicitly, we are more likely to heed such information when it expresses near unanimity, is delivered by perceived experts, or by people we think of as like us or important to us (see, for example, Aronson 2004). Achieving conformance through less direct means can draw on even more subtle features of our psychology. Of these, perhaps the most impressive is the power of mere familiarity over preference.

Kunst-Wilson and Zajonc (1980) famously showed that that familiarity of the most innocuous nature can produce a preference bias. Subjects were exposed for 1 millisecond to each member of a set of 10 irregular octagons. The 1 millisecond exposure length had previously been established to be too short to produce anything better than chance results in a recognition memory test in which the subjects were asked to discriminate the previously exposed stimuli from novel ones. However, when subjects were asked to make pair-wise preference judgments between random pairings of the previously exposed stimuli and novel ones (both of which were irregular octagons), a statistically significant preference for the previously exposed stimuli was demonstrated. (16 of 24 subjects preferred the old stimuli to the novel ones but only 5 of the 24 recognized them.) Kunst-Wilson and Zajonc used this data to argue that we have the capacity to make judgments noncognitively and their research has become central to a debate about whether cognitive representation is a necessary condition for emotional responses to stimuli. The conclusion I want to draw is different and twofold: first, familiarity affects preferences. But more importantly, familiarity starts much further from home than we might think, even in the domain of what is, for all intents and purposes, the unfamiliar. As a result, any pair-wise choice situation may be directly subject to preference effects without our even knowing it. So, while the behavior of others conveys information on which we can base rational choice, it also creates familiarity that can engender conformity as an accidental consequence of the alignment of preferences.

IV

I began this chapter assuming that not smoking was a rational choice about risk based on a cost-benefit analysis, and asked why the same does not hold for doing what we need to avoid climate change, even if we bracket the complications of collectivity and epistemic ignorance. But the lesson of what I have argued for is that this is the wrong way to set up the problem. The decision of a young person to stop smoking looks less like a rational cost-benefit analysis once we undermine the assumption of the self now and the self of the future as in some straightforward way the same self. Foregoing the pleasure of smoking now for the benefit of a later self looks less like self-interest and more like altruism. But a policy based on this assumes an individual decision is being made that balances risks and benefits even if the benefits inure to others. The lesson of successful antismoking campaigns is that such an analysis is far too individualistic. Giving up smoking looks like an individual rational decision after the fact, especially as an older person looks back at a decision made by a younger person (his former self). But in reality, individual smoking cessation was the end point of a social process that was collective at the outset. (The history of it is admirably chronicled in Fritschler and Rudder 2007.) More importantly, it was a process in which the terms of the public campaign to win individual support and the grounds for the collective policy ran on totally different tracks. There is no reason the same should not be true when it comes to climate change. The basis for our concern for regulating greenhouse gasses and the basis for which we argue for public support for such regulation need not be the same. And yet there is something disquieting about this conclusion, for what it does is to rob us of agency, both psychologically and morally. So perhaps we should not be so quick to embrace it and succumb to the temptation to the prescriptions that follow from it.

But, that said, what about the epistemic problem? When it comes to smoking, as we saw earlier, we know the risks in exquisite detail. Of course it is here that the contrast when it comes to climate change risk estimates could not be starker. And that brings us back to the question of how then to proceed when it comes to risk assessment for climate policy, as opposed to how to win acquiescence to that policy.

References

Aronson, Elliot 2004. *The Social Animal*, New York: Worth.

Centers for Disease Control and Prevention 2002. "Annual Smoking-Attributable Mortality, Years of Potential Life Lost, and Productivity Losses – United States, 1995–1999," *Morbidity and Mortality Weekly Report*, 51 (14): 300–303, www.cdc.gov/mmwr/preview/mmwrhtml/mm5114a2.htm, accessed June 2nd, 2011.

Farrelly, Matthew, Jeff Niederdeppe and J. Yarsevich 2003. "Youth Tobacco Prevention Mass Media Campaigns: Past, Present, and Future Directions," *Tobacco Control*, 12 (Suppl. I): i35–i47.

Farrelly, Matthew, Kevin C. Davis, M. Lyndon Haviland, Peter Messeri and Cheryl G. Healton 2005. "Evidence of a Dose–Response Relationship Between "truth" Antismoking Ads and Youth Smoking Prevalence," *American Journal of Public Health*, 95 (3): 425–431.

Fischhoff, Baruch, Paul Slovic, Sarah Lichtenstein, Stephen Read and Barbara Combs 1978. "How Safe is Safe Enough? A Psychometric Study of Attitudes Towards Technological Risks and Benefits," *Policy Sciences*, 9: 127–152.

Fritschler, A. Lee and Catherine Rudder 2007. *Smoking and Politics: Bureaucracy Centered Policymaking*, Upper Saddle River NJ: Pearson Education.

Goldman, Lisa and Stanton Glantz 1998. "Evaluation of Antismoking Advertising Campaigns," *Journal of the American Medical Association*, 279: 772–777.

Gore, Albert 2006. *An Inconvenient Truth*, Emmaus, PA: Rodale.

Hansen, James, Mki. Sato and R. Ruedy 2012. "Perception of Climate Change," *Proceedings of the National Academy of Science*, 109: 14726–14727.

Hu, Teh-Wei, Hai-Yen Sung and Theodore Keeler 1995. "Reducing Cigarette Consumption in California: Tobacco Taxes vs. an Anti-Smoking Media Campaign," *American Journal of Public Health*, 85: 1218–122.

Kunst-Wilson, William and Robert Zajonc 1980. "Affective Discrimination of Stimuli That Cannot Be Recognized," *Science*, 207: 557–558.

Moser, Susanne and Lisa Dilling 2004. "Making Climate Hot: Communicating the Urgency and Challenge of Global Climate Change," *Environment*, 46: 32–46.

Moser, Susanne 2007. "More Bad News: The Risk of Neglecting Emotional Responses to Climate Change Information," in Susanne Moser and Lisa Dilling (eds.), *Creating a Climate for Change*, New York: Cambridge University Press.

Parfit, Derek 1984. *Reasons and Persons*, Oxford: Oxford University Press.

Rabe, Barry and Christopher Borick 2011. *The Climate of Belief: American Public Opinion on Climate Change*, www.brookings.edu/papers/2010/01_climate_rabe_borick.aspx, accessed May 25th, 2011.

Slovic, Paul, Baruch Fischhoff and Sarah Lichtenstein 1980. "Facts and Fears: Understanding Perceived Risk," in R. C. Schwing and W. A. Alberts, Jr. (eds.), *Societal Risk Assessment: How Safe is Safe Enough?* New York: Plenum, 181–214.

Slovic, Paul, Baruch Fischhoff and Sarah Lichtenstein 1986. "The Psychometric Study of Risk Perception," in Vincent Covello, Joshua Menkes and Jeryl Mumpower (eds.), *Risk Evaluation and Management*, New York: Plenum, 3–24.

Swim, Janet, Susan Clayton, Thomas Doherty, Robert Gifford, George Howard, Joseph Reser, Paul Stern and Elke Weber 2009. *Psychology and Global Climate Change: Addressing a multi-faceted Phenomenon and Set of Challenges. A Report by the American Psychological Association's Task Force on the Interface between Psychology and Global Climate Change*, www.apa.org/science/about/publications/climate-change.aspx, accessed May 25th, 2011.

U.S. Department of Health and Human Services 2001. *Women and Smoking: A Report of the Surgeon General*. Rockville, MD: U.S. Department of Health and Human Services, Public Health Service, Office of the Surgeon General, www.cdc.gov/tobacco/data_statistics/sgr/2001/index.htm, accessed March 11th, 2011.

Woloshin, Steven, Lisa Schwartz and H. Gilbert Welch 2008. "The Risk of Death by Age, Sex, and Smoking Status in the United States: Putting Health Risks in Context," *Journal of the National Cancer Institute*, 100: 845–853.

2

THE COST OF EXTINCTION

I

A decision under uncertainty is a decision in which we can identify the alternative outcomes *and* attach probabilities to them. In a decision under ignorance all we have available to us are the alternatives but no basis on which to assign probabilities to them. Decision making under ignorance seems like an option of last resort, and one way to avoid it is to try one's best to assign probabilities to the outcomes under consideration. The challenges in trying to do that for climate change run up against the limits of both climate science and economics, but do so for different reasons. The problem for climate science is that the empirical record of data is too sparse to extrapolate from that record with much confidence. On the other hand the problem for economic theory is that the further in the future we try to project, the less confidence we can have about the applicability of our models. Moreover, both of these limitations are amplified by another consideration.

As I contemplate whether to buy long-term care insurance, the only open issue is the likelihood that I will need care and if so, for how long. The severity of the need does not enter into the picture, since the payout is fixed and barely covers the minimum daily costs for any kind of such care. But climate change is quite different in this regard. Looking to the future, we need to estimate what anthropogenic greenhouse gas output will be for a time slice. From that we need to estimate what net degree of warming will result. ("Net" because we also need to estimate the effects of countervailing human activities that have a cooling effect, like sulfur pollution.) From there, we need to estimate the damage such warming will produce and then go on to attach a cost to that damage estimate. Then we need to repeat the same procedure for both other effects of CO_2 output (like ocean acidification) as well as for other greenhouse gasses (like methane). Focusing just on CO_2, for any estimated level of output, there will be a probability distribution of the associated

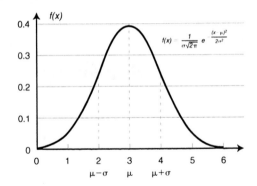

FIGURE 2.1 A Normal Distribution

costs. Given the state of our ignorance, especially the further we go out in time and up in CO_2 levels, outcomes may range from minimal to catastrophic. Let us assume that, all other things being equal, it is appropriate to represent this range as a normal distribution centered on our best estimate of the damage as in Figure 2.1. where

> e = constant
> μ = mean
> σ = standard deviation

In a normal distribution one standard deviation includes about 68% of the probability distribution, two standard deviations, 95%, and three, 99.7%. So, for any level of output, we can derive an expression representing the total estimated damage that is a function of each of the differing levels of estimated damage and their probability for each time slice. And then we can do the same for different time slices.

All of that is true for any prospective increased level of CO_2, but of course, we don't know what the level of CO_2 will be in the future absent global regulation that we currently lack; although all other things being equal, it is plausible to think that the higher the levels of CO_2, the more this distribution will shift to the right.

Missing in this epistemic fog is the valuational question of how much future damages should be discounted.

Of course all of these imponderables are more manageable the closer we choose to stay to the here and now. So it is all the more impressive that the canonical study of the cost of climate change, *The Stern Review* (Stern 2007), models estimate costs through the end of the 22nd century. Still it is important to keep in mind that the effects of increased CO_2 in the atmosphere are expected to be much more long lived, keeping in mind the warming of the oceans – on the order of millennia rather than centuries (see Archer 2009). That said, let's focus on a more limited time horizon and see what we can learn.

II

Let x be your best estimate of the level of the increase CO_2 in the atmosphere over year 2000 levels by the end of the next century. Let y be your best estimate of the net cost of that increase. Then, even if you are right about x, you may be wrong about y. So quite aside from the problem of getting x right, our challenge is to consider the range of values that y may take and the likelihood of it taking on any one of those values. This is not a standard problem of inferential statistics for the following reason. Consider the case of smoking again. Suppose I want to know how many years my life can be expected to be foreshortened (over the life expectancy of a non-smoker my age) if I smoked a pack a day for 45 years starting at age 20. The inference is both easy and reliable to make because the sample size of people who smoke is so large. Using the subset of those who have smoked a pack a day for 45 years, starting at age 20 and of those that are now dead, I can generate a probability density function, a distribution of the sample in terms of the number for foreshortened life expectancy. From there I can come up with an overall figure that is a function of those numbers and their probability. The obvious problem with applying the same process to come up with a figure for the value of y when it comes to the costs of CO_2 emissions is that we lack a ready to hand sample. What we do have however are economic models and perhaps those can be used to produce a probability distribution *based on the number of studies* for each value of y. But there is reason to be wary of the seeming parallelism here. Instances of smokers can be credibly accorded the same evidential weight, but not all studies deserve to be treated with the same degree of respect. Moreover, smokers may be plausibly treated as independent of each other. But studies often have incestuous pedigrees that make it hard to delimit their number. Still, thankfully, studies are all we have to guide us.

We face the same limitations in many other areas where we want to extend cost-benefit analysis to include worst-case scenarios that we have not had to face, the most obvious of which involve a complete core meltdown in a nuclear plant, notwithstanding the three (limited) accidents that have occurred to date. But there is an important difference between modeling the cost of a worst case nuclear plant disaster and our challenge. Not only is the modeling of a nuclear plant disaster relatively simple, but modeling it is of a piece with modeling other scenarios for which we do have (regrettably) actual empirical instances. That is to say, the model for a nuclear disaster is an extension of the model for the observed ranges of a plant's normal functioning along with models of the nuclear accidents to date. The challenge of modeling the cost of climate change is that the more we go out in time, and the more we focus on extreme possibilities, the more we are in terra incognita. Add to that fact that much of what drives this limitation is the scope of ambition of what we are trying to do. For after all, a model that will tell us how much damage our actions may cause to the world requires that we have a model of the whole world. Modeling this to produce a damage function is not just an uncertain enterprise. It also suffers from second order uncertainty. But even worse, the second order uncertainty here is irredeemable.

III

Martin Weitzman (2009, 2011) has argued that second order uncertainty like this produces a t-distribution that has a lower peak but fatter tails when compared to a normal distribution, as shown in Figure 2.2.

In our case, a fatter right hand tail is indicative of a higher likelihood of worse than the expected scenario for any level of CO_2 as compared to a normal distribution. But putting it this way understates just how radical Weitzman's view is. For his argument is that, when combined with a plausible level of risk aversion, the expected value of such a t-distribution has infinite disutility! Infinite disutility presents a problem for rational choice theory, but let's ignore this embarrassment for now. Weitzman is not asserting that the possibility of a catastrophic outcome poses infinite disutility despite its uncertainty. Instead he is arguing that it is the uncertainty itself (together with risk aversion) that produces the infinite disutility. How can this be?

Consider a scientist who plans to go into the field to collect samples to study the size of a particular ant species. Again, all other things being equal, it is reasonable to assume there will be a normal distribution around a mean. As such, most of the observations will congregate around that mean and deviations far from that mean will be observed with lower frequency. That means that it will be possible to get an estimate of the mean without knowing the variance of the distribution (that is, the probability distribution) of the population. But absent knowing the variance, a normal distribution about a mean may be realized in a wide variety of ways at the tails as shown in Figure 2.3.

However, over time as the sample size is expanded, more data on outliers will allow the estimate of the probability distribution to improve. But now suppose it is not possible to sample for outliers. Not because they don't exist, but because of some limitations the scientist faces. Perhaps the budget is too limited to gain enough data. Then uncertainty about the probability distribution will not diminish. So confidence about the mean value of the sample can coexist with uncertainty about the tails. But that uncertainty can itself be expressed as a probability distribution, a probability distribution that ranges over the underlying (normal) probability distributions.

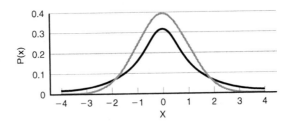

FIGURE 2.2 Normal and T Distributions

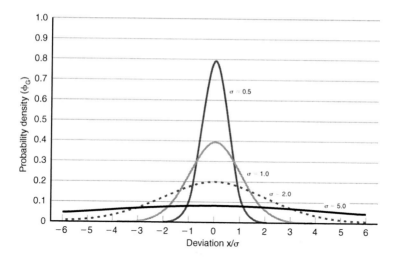

FIGURE 2.3 A Family of Normal Distributions

Now in the case of our ant scientist, funds may come in to allow more research to reduce this second order variance as our confidence in which is the correct underlying probability distribution improves. But in the case of climate, money is not the issue. Weitzman's argument is that the limitations of our models, and the lack of an empirical record, give us little guidance about what the right probability distribution is the further out we go along the right hand tail. And hopefully no actual data is going to be available to us. On that basis he argues that this *second order distribution* will have the "fat" tail of a t-distribution.

Any tail, normal or fat, is asymptotic, trending downward toward a limit value but never reaching it. But a fat tail trends down more slowly than a normal tail. Sitting here in the present, our conundrum is to decide how much to discount future uncertainty in which there is a chance that wealth (y) may fall to zero. How much would we be willing to pay now to avoid the chance of that outcome in the future? Assume (quite conservatively) that our aversion to risk (η) is constant. (That is, it does not increase the further we go out on the right hand tail toward worse and worse outcomes.) Then, even with that conservative assumption, when we calculate the discount factor between the present and a time slice in the future, the willingness to pay now (E(M)) becomes infinite, which is to say, the disutility associated with the outcome we are willing to pay to avoid, becomes infinite. The reason why is that in the canonical formulation for the stochastic discount factor:

$$E(M) = \int_{-\infty}^{\infty} e^{-\eta y} f(y) dy$$

f(y) does not fall at a fast enough rate in a t-distribution to offset the (exponential) rate at which $e^{-\eta y}$ grows.

Given the uncertainty of both climate change science and economics as we look to the future, Weitzman's argument is profoundly seductive. In his hands all of this uncertainty produces a result that seems to remove any uncertainty about the pressing need for climate action. For notice, all manner of discounting, be it based on knowledge, future technological innovation, or wealth and the value of money, gets swamped out by infinite disutility. And so worrying about the size of those discounts seems moot.

But is this a result arrived at by honest toil or theft (to echo Bertrand Russell (1919, 17) in a different context)? Weitzman's result rides on assumptions about the underlying probability distributions (as normal), risk aversion (as constant), and the second-order distribution (as a t-distribution). Now while there is no reason to question the assumption that the underlying distribution is normal, change any one of these and the argument can be undermined. Critics of Weitzman have focused on the last of them, arguing that a t-distribution unjustifiably assumes we have a total state of ignorance about which underlying distributions might be the right ones. An alternative line of attack is to argue, as some economists have, that risk aversion is decreasing rather than constant. (For both, see especially Nordhaus 2009.) Neither of these seem like plausible lines of attack to me. Whatever the reliability of our economic models may be, the damage function we are dealing with relies on climate models that are widely acknowledged to be uncertain beyond a limited range of experience, which should sap confidence the further out we move in the range of the right hand tail. Moreover, far from substituting constant risk aversion for diminishing risk aversion, it seems reasonable to think that risk aversion should *increase* the further we move out on the right hand tail.

IV

We have been examining arguments for and against the infinite disutility of risking climate change in which uncertainty does the work. One thing that is noteworthy about this argument is how little work the *subject* of that uncertainty actually does. In his work Weitzman is certainly at pains to show how his econometric argument is consistent with the dire projections of climate scientists, but those projections themselves play no role in the econometric argument itself. As such, if Weitzman's argument is sound, its soundness is insensitive to *any* level of CO_2 above historic levels to the extent that the probability distribution for damages at that level has a fat right hand tail. In that sense, Weitzman's argument, even when restricted to climate change, is remarkably promiscuous in producing its result with infinite disutility, and all the more so given that the result is arrived at with a time horizon restricted to just 200 years. What would it take to arrive at the same result without relying on the work that the fat tail of uncertainty itself does in Weitzman's argument? In particular, could we arrive at the same result with the disutility of a particular outcome along the tail doing all the work?

It is important to note that this question only makes sense if a particular worst case outcome has infinite costs associated with it. That is very different from standard approaches. For example, Nordaus and Boyer (2000, 90–91) treat a catastrophic outcome as a global 30% reduction in GDP (ranging from a low of 22.1% for some regions and 44.2% for others). Needless to say, outcomes with infinite costs are insensitive to the probability of them coming about. As long as the chance of such an outcome is greater than zero, the costs will remain infinite. But what would render the costs infinite? At first blush the extinction of Homo sapiens might seem to do the trick, for extinction denies future generations existence they might reasonably be expected to have had. Let's grant this for now. Is there any reason to think that climate change could have such extreme consequences?

As we saw earlier, James Hansen (2008) thinks there is. In his Venus scenario, increasing water vapor as the planet warms drives things and (unlike cooling) there is no naturally occurring reversing process in the case of runaway heating, resulting in mass extinction. But what level of global warming might trigger such a scenario? (Here and throughout I follow Hansen in discussing only CO_2 levels rather than the equivalently higher levels used when discussing CO_2 output and other greenhouse gasses that by convention are indicated as CO_2e.) Hansen et al.'s (2005) modeling suggests a figure of 10 to 20 watts per square meter (W/m^2) of added forcing above current levels. What does the lower bound of this mean for CO_2 output levels? Atmospheric CO_2 is now roughly 125 ppm (parts per million) higher than it was during pre-industrial times (at 400 ppm) and the associated increased radiative forcing is 1.88 W/m^2. An additional 10 W/m^2 would be equivalent to adding 640 ppm of CO_2 to the atmosphere for a total of roughly 1040 ppm. As Ben Kravitz pointed out (in a private communication) converting this into a temperature isn't so straightforward and depends on what time scale we use. In the short run, the associated rise is only (!) 0.81 °C for +1 W/m^2. But Hansen's runaway scenario includes the slow feedbacks over longer time scales for which the rise might be 2.5 °C for +1 W/m^2 as a result of the cumulative effects of ocean warming and ice sheet melting. It may seem that 1040 ppm of CO_2 is far away. But on a business as usual model (RCP8.5) we get close to it by the end of the century in the IPCC summary of projections as seen in Figure 2.4.

Moreover, the likelihood of such a worst-case scenario doesn't really matter, as long as it has a probability greater than zero, given the costs if it were to occur, which we have taken to be infinite, assuming the end of all human life.

On that basis we should spend whatever it takes to avoid these consequences, and with that all of the calculations of probability and net present value become irrelevant. (My colleague Robert Kopp has pointed out that you can get to conditions hostile to life long before you reach the levels Hansen is concerned about, and you can do so without having to posit runaway conditions as Hansen does. But if you want to consider a scenario in which *all* life is destroyed, including deep sea bacteria and tube worms, Hansen's is a better worst case scenario to worry about, even if it has a lower probability than less extreme scenarios.)

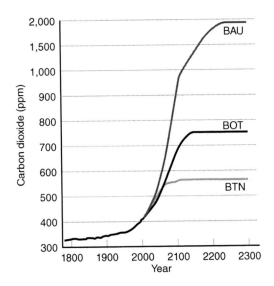

FIGURE 2.4 Projected Atmospheric CO_2 Concentrations by Scenario (IPCC 2013, Chapter 1, 148, Box 1.1, Figure 2.)

Not so fast! First, by parallel argument we ought to do the same for other threats that have a nonzero chance of destroying us. But we don't. Second, even if we did, there are too many other threats for us to protect ourselves against them all.

If a large enough (killer) asteroid were to hit Earth it could destroy all life. Following Chapman and Morrison (1994), suppose the chances of that are assumed to be 10^{-8} per year. But as Nordhaus points out (2009, 13), we currently spend only $4 million a year tracking such asteroids although an outlay of about $1 billion per year "could reduce the probability of impact by at least 90 percent, but this sum is apparently not worth the avoided risk." In practice then, argues Nordhaus, if we assume this to be the "outcome of reasoned choice," we don't display "highly risk-averse utility" (2009, 14). But why should that descriptive finding have prescriptive implications? After all we are interested here in how we should act, not how we do in fact act in the face of climate change. Moreover, even if we accept Nordhaus' assumption that how we do act is the product of reasoned outcome for the purposes of argument, this distinction still holds. How we do act can be reasoned but based on valuations for outcomes that are misguided from a prescriptive point of view.

Perhaps the problem is different. Even if we ought to be willing to spend whatever it takes to avoid the risk of extinction through climate change, as long as doing so does not itself create a risk of extinction, what if that is not the only source of risk of extinction? Both Sunstein (2007, 172) and Nordhaus (2009) argue that there are plenty of other sources for which our uncertainty does not allow us to rule out the possibility of catastrophic outcomes. As Nordhaus (2009, 14) puts it: "Example

[sic] would include biotechnology, strangelets, runaway computer systems, nuclear proliferation, rogue weeds and bugs, nanotechnology, emerging tropical diseases, alien invaders, asteroids, enslavement by advanced robots, and so on." But even if these do all pose a risk of catastrophe, do they pose a risk of extinction?

"Well," you might say, "anything is *possible*." We can never know for sure that these might not pose a nonzero risk. And not just them. Possibility is a permissive notion. Indeed, if we need to worry about anything that is a possible cause of extinction, understood as something we can't rule it out as a possible cause, the list will be endless. Assume we only need to worry about what is possible relative to our physical laws. Still, we need to worry about the veracity of our knowledge of those laws. Lots of things may go wrong because our knowledge is not as robust as we thought. We may assume nature is benign and fail to intervene to avoid catastrophes when we could have. On the other hand, we may erroneously assume our actions are benign when we intervene in nature. As we look around, danger may lurk in both actions we fail to take and actions we might take. But if this still looks like a prescription for indecision, it is only because we are implicitly privileging the status quo. But we have no basis to be *sure* that conducting our affairs as we do is risk free itself. Either way, when it comes to the risks posed by nature, all we can hope for is to be guided by our best theories and the empirical record to both assess those risks and the risks of actions to avert them. Of course few of the risks nature poses are amenable to offsetting actions by us, and even where they are available to us, the risks of those actions may outweigh the risks posed by nature.

But not every risk posed by nature is of interest here, for our question is: based on our best theories and the empirical record, what risks does nature pose to *extinction* that might we avert without creating an equal or greater risk? Bracketing considerations of climate change, surprisingly the answer I think is at most one beyond killer asteroids: namely, viruses. But to even make the case that viruses pose this risk requires taking some scientific liberties.

Twenty-five million people have been killed by HIV and 75 million people were killed by the 1918 flu. Horrendous as these pandemics have been, they fall very far short of posing a risk to human existence. HIV has killed .5% of the current world population while the 1918 flu killed 3% of the world population of the time. Evidence of HIV in humans preexists the current pandemic and supports the hypothesis that it previously crossed over from a reservoir in other animals without spreading far in the human population. The key difference between those crossovers and the current pandemic was changes in human mobility.

That said, our knowledge of viruses is in its infancy and the vast proportion of them have yet to be identified. Zimmer (2011) reports that only 10% of viral species in the lungs of human subjects have been identified. Assuming that a similar rate of ignorance holds for our knowledge of viruses throughout nature, it is a safe bet that there exist large reservoirs of viruses that will be lethal if they jump to humans. Moreover, human population growth, deforestation, and rising consumption of primates increase the opportunities for such cross-overs to occur. But as Zimmer (2011) points out, the Ebola virus spreads easily but kills too quickly to be able to gain a foothold. On the other hand, the HIV virus kills slowly, but does not spread easily enough to threaten the whole species, even absent any treatment. For a virus to pose a risk of human extinction, it would have to combine ease of infection with high but slow-acting lethality. Whether there exists a reservoir for such a virus can't be ruled out given our current ignorance about viruses. But even if it could be ruled out, nothing can rule out the possibility of a mutation that exhibits these traits.

In the case of asteroids, we have no guarantee that we could avert catastrophe, but we could develop the technology (if we don't have it now) to diminish the chances of such a catastrophe (see Committee to Review Near-Earth Object Surveys and Hazard Mitigation Strategies; National Research Council 2010). In the case of viruses, the same is true, but, even without an increase in our knowledge and technology, we could choose to implement land management policies to diminish the opportunities for viruses to cross over from existing reservoirs into the human population. Asteroids and (perhaps) viruses are not the only risks nature poses to our continued existence. Consider a large number of simultaneous volcanic eruptions that produce enough emissions to block the sun for many years. But such volcanic eruptions are the kinds of risks that we are powerless to change.

Of course the real risk to our existence comes not so much from nature but from our own actions that we are not powerless to change. The anthropocene is such an example par excellence. But can we really be *sure* that other of our other actions don't pose at least *some* risk of extinction. Notice that the concern here is not a risk of generalized catastrophe à la Sunstein and Nordhaus, but only of catastrophe large enough to precipitate extinction and that therefore ought also to be eschewed. If we are to worry about the extinction risk of greenhouse gases, then why not also worry about the risk of genetic manipulation, or nuclear energy, or particle accelerators, or . . . the list would seem endless since it is surely impossible to rule out *all* risk of extinction.

"Better safe than sorry" you may say echoing the most extreme form of the (do nothing unless you can be sure you do no harm) precautionary principle. On this view, activities like genetic engineering, nuclear energy and the like may have *some* benefit, but any such benefit is overwhelmed by the possibility of the (infinite) cost of extinction. Let's divide the list of potentially risky actions into two: those for which we have a basis in knowledge in the empirical record to believe they might cause our extinction (however small the chances) and those for which we lack such a basis but can't rule out all risk. Suppose we eschew the former. The worry is with how to treat this second group. The proponent of the most conservative precautionary principle will argue that we should treat these as we treat those about which we know there is risk and eschew them as well, acting only on those cases in which we can be sure there is no risk. Taken as a general principle this version of the precautionary principle can't work. The reach of our knowledge is too limited to offer the requisite level of certainty that the principle demands when it comes to eliminating the chance of any adverse consequences of our actions. But remember here we are not talking about risk in general but risk of extinction. There may be many actions for which I can't be sure they won't cause some harm or more harm than good, but for which I can be confident that they won't cause extinction. Consider the most salient of these: genetically modified foods. Many people worry that the introduction of genetically modified crop strains may have unintended consequences for other species, in their immediate ecological milieu or by propagating more widely than intended. Such propagation might pose a threat to some species, especially if their members are not widely distributed. But nobody has suggested that human extinction is a worst-case scenario for genetically modified crops. And with good reason, our knowledge of natural selection and the empirical history of selective breeding underwrite our confidence. To precipitate extinction of a species as widespread and entrenched as ours by our own actions (as opposed to those of nature) is not that easy. It would require actions that have global impact – as extreme as climate change (à la Hansen scenario). Are there other such actions? The obvious ones are global thermonuclear exchange and biological warfare. The less obvious one involves a change in the loci of viruses, for which, as we discussed above, we may play a precipitating role.

I have been arguing that, although not sui generis as a threat to our existence, climate is only one among relatively few threats of that kind. As such, the idea that, if we act to avoid one of these threats, we ought to act on them all, posed as a reductio argument, loses its force. But what about the risks of Hansen's scenario at lower levels elevated CO_2 output than that envisioned by Hansen? Since we don't know if our models of climate sensitivity are accurate, carbon output at lower rates than he considers *might* produce the warming his model is concerned about. The likelihood would seem to be a lot less than at the level of output that Hansen considers, but our ignorance does not justify setting it at zero for levels outside our own experience. Does that mean *no* level is safe?

If not, then ought we not to forgo any economic output now that could pose a danger of extinction in the future? Any? Not if doing so would itself create a state of affairs that would itself produce extinction. But anything other than that would seem to be fair game. The most extreme alternative under this cautious prescription would be to sustain no more than a limited number of people at no more than a subsistence level. Limited but large enough to sustain the species. That may seem preposterous but not in itself irrational, at least if we make decisions with the interests of future generations in mind. And if it seems more draconian than what even advocates of a simpler life (like Bill McKibben 2006) call for, keep in mind that this is the most extreme alternative. A more realistic alternative would be to stabilize atmospheric carbon no higher than the pre-industrial level of (roughly) 275 ppm, although Hansen himself has advocated a target of 350 ppm (Hansen et al. 2008). Here both theory and the empirical record help again. For we have a record of climate at pre-industrial levels of carbon output as well as ice core data both of which support the conclusion that levels in the 350 ppm CO_2 range or lower pose no risk of extinction, whatever other risks they pose. That said, be it 350 ppm or 275 ppm, we are already far above both of these levels, so such a goal would mean we would either need to reduce our carbon output to a level below the rate at which CO_2

comes out of the atmosphere through natural processes to reach a steady state at the requisite level, or develop a way to remove it from the atmosphere ourselves. (I take this question up in Chapter Ten.)

Contrast my view to Nordhaus (2013, 113). Nordhaus would have us decide on a target for limiting warming based on a cost benefit analysis that optimizes "the twin objectives of preserving our environment for the future while economizing on losses in the living standards along the way." But what if we are unsure about where catastrophic thresholds lie? Nordhaus (2013, 216–217) thinks we can easily incorporate a precautionary principle into his calculus to avoid such thresholds by setting a limit below them all. But he thinks this can only be plausible if "there are only a limited number of potential cliff in the damage function, and further it is not ruinously expensive to avoid them all." But why should the number of unknown cliffs make a difference – as long as the "ruinous" cost of avoiding them is not worse than exposing ourselves to them? If there is a difference here (as we saw before) it revolves around what counts as a catastrophic outcome. For Nordhaus it is a significant loss of global income while for me it is extinction. So for him, unlike me, what counts as a catastrophic outcome is one we can recover from in time.

V

The argument so far has treated the cost of extinction as infinite, be it on the basis of Weitzman's argument or Hansen's. That is what obviates the need to worry about how much it would cost to avoid extinction. As long as the costs to avoid extinction are finite, no amount of discounting of the costs of extinction will make the cost of avoiding extinction greater. But there are two very different reasons to be suspicious about treating the cost of extinction as infinite. One derives from considerations of rationality and the other from considerations about our long-term prospects as a species notwithstanding climate change.

"There is a natural tendency to scoff at economic models that yield infinite outcomes" writes Weitzman (2011, 17) and with good reason. Unbounded values undermine (mathematically) well behaved properties of utility functions and with that, produce results inconsistent with the axioms of rational choice. And although Weitzman thinks infinite disutility "is trying to tell us something important" (2011, 17), he is nonetheless willing to limit disutility to avoid mathematical anarchy. Even without settling the question of just where to place the limit, this move straightforwardly deprives us of the basis for the confidence that the cost of extinction will always outweigh the cost of avoiding it, however small the chances of it occurring. Now you might say, so much the worse for the concept of utility functions. But unfortunately, it is not just the requirements of utility functions that create the problem.

Until now we have also assumed that if we cause the extinction of our species, the costs are infinite because of the lives forgone. But is that correct? Even on the most optimistic scenario in which asteroids and viruses are tamed, the sun will eventually burn out and engulf the Earth. That is bound to occur. So even if hundreds of millions of generations of our descendants might follow us, their number is not infinite and hence the costs of denying them existence are not infinite either, even if they are very, very large. We can save the ease of the argument that assumes infinite costs only by assuming some of our kind could relocate to a different solar system in time. And indeed some people, including Stephen Hawking, have advocated just such a strategy (BBC News 2006). But what if we don't rely on that assumption? Once the cost of extinction is treated as finite, can we be sure that avoiding the risk of it occurring is worth it? Without the crutch of infinite numbers, we need to ask what the probability of the lives foregone would be, the cost of those lives, and the net present value of those costs. Or must we? Even if the numbers are not infinite, are they so large that neither a small probability nor a large amount of discounting will make much difference? If so, we could treat the numbers as effectively infinite.

Assume the sun will burn out in 5 billion years. Assume four generations per century and that the population remains stable at 9 billion people. So the people who will follow us will be roughly $(5 \times 10^9) \times (9 \times 10^9) \times 0.04 = 18 \times 10^{17}$. Assume we treat these lives as all of equal value (in constant dollars). The value of a statistical life is typically based on the study of willingness of people to pay for risk reduction or the addition to wages necessary to get people to be willing to accept additional work related risk. (For example, in the case of the former, if people are willing to pay an additional $1,000 for safety options on a car that reduce the risk of death by 1 in 1,000 over the life of the vehicle, then the statistical value of life revealed is $1,000,000, the total amount spent per life saved. Similarly, in the case of the latter, if the wage differential between a regular construction job and one with an increased risk of death of 1 in 100 per annum is $10,000 per annum, then the value of statistical life revealed is also $1,000,000.) The value of statistical life varies from country to country. Indeed, valuation even varies across U.S. federal agencies, ranging between $1 million and $10 million (Robinson and Hammitt 2010).

Aiming close to the middle of this range, assume all lives are valued at $5,000,000 in constant dollars. So the cost of extinction is 9×10^{24} before we take into account the probability of it occurring or the discount rate. Let's assume the best, and with that, assume that Hansen's Venus scenario is unlikely with a chance of only 1 in 100,000,000, the same rate as we assumed for a killer asteroid hitting us in any one year (based on Chapman and Morrison 1994). That is to say, each year that our carbon output exceeds 350 ppm we run a nonzero risk of precipitating the Venus scenario even if it takes many thousands of years to unfold. During that period there will be more economic output and fewer people will be denied existence, but for all intents and purposes we can ignore this given the 5 billion year horizon in our calculation. Then the cost of extinction for any one year is reduced to 9×10^{16}. That is a lot larger than current annual world economic output of 74×10^{12}, let alone the output attributable to our exceeding 350 ppm of 54×10^{12}. (That level of CO_2 concentration was reached in 1988 when world GDP was 20.3×10^{12}.

See International Monetary Fund 2010 and National Oceanic and Atmospheric Administration 2011.)

Using this second figure, we can pose the ratio as between the economic gain and the potential loss in the form of betting odds. You are offered a chance for a level of economic gain of 54×10^{12} but in doing so there is a risk of causing extinction at a cost of 9×10^{24}. What would the odds of the latter need to be for the bet to be even? It would be $9 \times 10^{24} / 54 \times 10^{12}$. That is $1 : 1.7 \times 10^{11}$, which is considerably less than the $1 : 1 \times 10^{8}$ odds that we assigned as the likelihood.

But what about this consideration: the sun may burn out in 5 billion years but it will make life on earth hard to sustain long before then. Let's ask when that would have to happen to make the calculation come out differently. If we stick with the assigned risk, the same population and value of statistical life assumptions, then how many years would we need to assume before the sun makes life on earth impossible for a bet here to be even? We need to know the value of t in the expression:

$t/5 \times 10^{9} = 1 \times 10^{8}/1.7 \times 10^{11}$, which comes out to a little more than 3 million years!

Now of course that leaves out the discount rate. But it is not clear that we should set a discount rate here. For any year we want to know if the benefits of world output are "worth" the risk of precipitating extinction *in that same year* both measured in constant dollars (ignoring economic growth).

VI

A reservation about the argument so far. We have treated the costs of avoiding extinction as finite. If we chose to live on a strict carbon diet to avoid any further carbon output and thus limit CO2 at current level, we would have to change our lifestyles radically. World economic output would fall dramatically along with economic growth. Those losses would still be finite. But what about the loss of population as compared with what it would have been with higher output and growth? If avoiding climate change means lowering economic output and with it the population that could be sustained, we deny life to some who would have been born and their descendants. But even if those denied existence are themselves large in number, for any arbitrarily chosen point in time, they will be less in number that those denied existence by the extinction of the whole species.

I have been arguing that even if we don't treat future generations as infinite in number, if we take their welfare into account and take a worst case scenario about climate change seriously, prudence favors avoiding the risk of a Hansen type scenario. What about the "if"? After all, those that might come after us won't come if we destroy the planet. In what sense is there then a loss to them?

Is this question anything more than an exercise in philosophical sophistry? It turns out to be surprisingly difficult to ground the commonsense intuition that those that might have existed, but don't, lose out. But making sense of it is crucial to the argument so far. Along with it there comes another commonsense intuition that deserves consideration: the argument so far has been unabashedly anthropocentric. Making it less so would only bolster the virtues of prudence in our carbon output. But to make that case means talking not only about the interests of future generations of humans who might not be, but also the interests of other living things and, maybe, even "Nature" itself.

References

Archer, David 2009. *The Long Thaw*, Princeton: Princeton University Press.

BBC News 2006. "Move to New Planet, says Hawking," *BBC News*, http://news.bbc.co.uk/2/hi/uk_news/6158855.stm, accessed March 12th, 2012.

Chapman, Clark and David Morrison 1994. "Impacts on the Earth by Asteroids and Comets: Assessing the Hazard," *Nature*, 367: 33–40.

Committee to Review Near-Earth Object Surveys and Hazard Mitigation Strategies; National Research Council 2010. *Defending Planet Earth: Near-Earth Objects Surveys and Hazard Mitigation Strategies*, Washington DC: National Academies Press.

Hansen, James 2008. "Climate Threat to the Planet: Implications for Energy Policy and Intergenerational Justice," Lecture given Dec. 17 at the American Geophysical Union, San Francisco. (Slides posted at www.columbia.edu/~jeh1/presentations.shtml, accessed June 4th, 2011.)

Hansen, James, M. Sato, R. Ruedy, L. Nazarenko, A. Lacis, G. Schmidt, G. Russell, I. Aleinov, M. Bauer, S. Bauer, N. Bell, B. Cairns, V. Canuto, M. Chandler, Y. Cheng, A. Del Genio, G. Faluvegi, E. Fleming, A. Friend, T. Hall, C. Jackman, M. Kelley, N. Kiang, D. Koch, J. Lean, J. Lerner, K. Lo, S. Menon, R. Miller, P. Minnis, T. Novakov, V. Oinas, J. Perlwitz, D. Rind, A. Romanou, D. Shindell, P. Stone, S. Sun, N. Tausnev, D. Thresher, B. Wielicki, T. Wong, M. Yao and S. Zhang 2005. "Efficacy of Climate Forcings," *Journal of Geophysical Research: Atmospheres*, 110: D18104.

International Monetary Fund 2010. *World Economic Outlook Database*, www.imf.org/external/pubs/ft/weo/2010/02/weodata/download.aspx, accessed September 3rd, 2012.

IPCC 2013. Climate Change 2013: The Physical Science Basis, available at www.ipcc.ch/report/ar5/wg1/#.UmGB0FCkojU, accessed October 18th, 2013.

McKibben, Bill 2006. *The End of Nature*, New York: Random House.

National Oceanic & Atmospheric Administration 2011, *Up-to-date Weekly Average CO2 at Mauna Loa*, www.esrl.noaa.gov/gmd/ccgg/trends/weekly.html, accessed August 10th, 2011.

Nordhaus, William 2009. "An Analysis of the Dismal Theorem," http://cowles.econ.yale.edu/P/cd/d16b/d1686.pdf, accessed July 25th, 2013.

Nordhaus, William 2013. *The Climate Casino*, New Haven: Yale University Press.

Nordhaus, William and Joseph Boyer 2000. *Warming the World: Economic Models of Global Warming*, Cambridge: MIT Press.

Robinson, Lisa and James Hammitt 2010. *Regulatory Analysis: Current Issues and Challenges, Working Paper No. 4*, Jerusalem: Jerusalem Forum on Regulation & Governance, www.regulatory-analysis.com/robinson-hammitt-valuing-health.pdf, accessed August 10th, 2011.

Russell, Bertrand 1919. *Introduction to Mathematical Philosophy*, London: Allen and Urwin.

Stern, Nicholas 2007. *The Economics of Climate Change: The Stern Review*, Cambridge: Cambridge University Press.

Sunstein, Cass 2007. *Worst Case Scenarios*, Cambridge: Harvard University Press.

Weitzman, Martin 2009. "On Modeling and Interpreting the Economics of Catastrophic Climate Change," *Review of Economics and Statistics*, 91(1): 1–19.

Weitzman, Martin 2011. "Fat-Tailed Uncertainty in the Economics of Catastrophic Climate Change," *REEP Symposium on Fat Tails*, www.economics.harvard.edu/faculty/weitzman/files/REEP2011%2Bfat-tail.pdf, accessed June 21st, 2011.

Zimmer, Carl 2011. *Planet of Viruses*, Chicago: University of Chicago Press.

3

THE RIGHTS OF THOSE
WHO WILL NOT BE

I

Paul Hawken (1994, 139) admonishes us to "[l]eave the world better than you found it, take no more than you need, try not to harm life or the environment, make amends if you do." Let us assume we do the opposite. Then whom do we wrong? It takes a philosopher to find fault with the obvious answer of future generations.

Before the philosophers complicate things, that we owe something to future generations seems obvious if we concede that we owe something to those alive now when it comes to sharing scarce resources. Indeed it seems to follow directly from it unless you think time makes a difference. Your spaceship breaks down and you are forced into an emergency pod (a lifeboat in space) with nine others and a limited amount of food and water. How should you dole out the supplies? For most people, the obvious answer is that we should do so on an equal basis, all other things being equal. If you hold onto that idea for the purposes of argument (whether you agree with it or not), now imagine that the spaceship was able to travel in time and took on passengers, so some come from the future. At least by my lights, this leaves my intuitions about sharing the resources on an equal basis unchanged. Nor do they change on any alternative formula, say one based on need or on benefit. Of course, the spaceship is a metaphor for Earth, and the passengers are its inhabitants, both now and in the future, who must share its limited resources. Time travel helps here because it lets us put those in the future and those here now on an equal moral footing. Without it we may be queasy about embracing the idea that those who don't exist now have a claim on us now.

But do they? The intuition that they do not is that only people can have such claims on us, not potential people, and people located in the future are only potential people or possible people in the here and now. So we either need to allow that possible people do in fact have claims on us, or allow that people in the future can have claims that can reach back to times at which they don't exist. But the first alternative is an obvious non-starter, if only because of the infinite number of possible people. If we treat them all has having a claim as actual people, and share equally, everyone's share will approach zero. You might object that merely possible people don't need a share of anything, it is only those who are potential people now and will become actual people later who do, when that need is viewed atemporally. But that is just to embrace the second position. What makes that second position seem problematic is that it assumes certainty about the future now. And that is what we are doing in the time travel thought experiment. At most, what the thought experiment shows is that, if we assume certainty, then treating those future beings as having claims on us now seems straightforward (at least it does if we think it is straightforward that we have claims on each other in the first place). The problem arises when we confuse that with the more realistic state of affairs in which we don't know which possible people will become actual people in the future. In some cases we can just wait and see. When my mother died, she left money to be shared after 5 years on an equal basis among her grandchildren, both those alive at the time and any that were born within 5 years of her death. But such arrangements are not a luxury we have when it comes to fair dealing *all* the people who will live in the future.

Still, why does it matter if we don't know who they are as long as we know their number? And while we may not know their exact number, as outlined in the last chapter, we can make a rough estimate and that is enough to listen to Hawken: we should strive to leave things better (or, if not, at least no worse) than the way we found them for those who will follow us. A corollary of that would seem to be this: if we don't listen to Hawken we wrong those who will follow us.

II

But this picture is too simple, for it ignores the fact that what we do now not only affects the quality of life of those who follow but also *who they are* (see Parfit 1984). A couple that postpones having a child for a few years may rationalize doing so on the grounds that "the child will have a better life." But this glosses over the fact that that child will not have the same identity as the child that would have been born had the couple not postponed reproduction. Likewise, the child of a couple that decides not to postpone reproduction may complain that "you could have given me a better life if you had only waited." But the child that would have had the better life would not have been the complainant.

Now consider a more extreme variant of this problem. Suppose I contaminate the drinking water in our county but the effects are not apparent until it is too late. Suppose the contamination has two consequences. It makes it harder for women to conceive, and when they do, they bear children with birth defects. Suppose that, but for the contamination, a couple would have conceived within the first 3 months of trying to get pregnant resulting in a healthy infant – Henry. Instead it takes them six months longer to conceive, resulting in a damaged infant – Henrietta. But for the contamination, Henry would have been born, conceived of a particular ovum and sperm. Because of the contamination that conception does not occur. If it had, the couple would not have gone on to conceive Henrietta. My actions produce Henrietta rather than Henry, and my actions render her life of lower quality than what his life would have been. Yet Henrietta can have no complaint, but for my actions she would not have existed. And Henry never existed, so in what sense did I cause him a wrong? But surely I did do wrong, and capturing that is the philosophical challenge addressed in the existing literature on future generations.

The locus classicus of an answer to this problem is Parfit's proposal that is formulated as a counterfactual:

> If in either of two outcomes the same number of people would ever live, it would be bad if those who live are worse off, or have a lower quality of life, than those who would have lived.
>
> (Parfit 1984, 360)

For Parfit, this formulation is a way station in the search for a formulation that can also accommodate contrasting outcomes in which the numbers are different. That is easier said than done, and it too depends on a counterfactual contrast, but for our purposes, the simpler formulation above will do. For here, we only have two people. Neither is wronged by what I do, nor is the "bad," in the sense that "it would be bad," bad for either of them.

Talk of wrongs here need not necessarily imply rights, and Parfit's formulation discharges such rights talk in favor of an account of wrongs based on beneficence. Henry never was and so, unless potential people have rights, he has none. On the other hand, Henrietta is a bona fide rights bearer. But suppose, to echo Parfit (1984, 364), she has a right to a good start in life. Then, "even if this child has this right, it could not have been fulfilled." If someone had been born with a good start, it would not have been Henrietta. So this prescription can't be a matter of a violation of some right of hers. Cashing the wrong I do in terms of beneficence sidesteps these problems. But then we need to ask how promiscuous this notion of beneficence is. In particular, suppose that producing more people could increase the total amount of well-being in the world. Must this not fall within the purview of the demands of beneficence? At first blush it might seem tempting to follow Narveson (1967), and say, "it is not good that people exist because their lives contain happiness. Rather, happiness is good because it is good for people" as Parfit pithily characterizes the position (Parfit 1984, 394).

Indeed, I find it hard to take anything but a stance of indifference toward the opportunity to increase total happiness in the world by increasing the number of happy people in the following sense: assume n is the number of people who now exist and will exist, and x is the sum total of their happiness. Now assume I have it in my power to add one more person to n who would not otherwise exist and that their life would be happy for them, but not add or detract from the happiness of others. My indifference here is grounded in part on a thought experiment due to Christoph Fehige (1998, 513–514). Suppose Kate is indifferent about the color of a particular tree, but that we paint it red and simultaneously give her a pill that creates a desire in her that it be red. Is she any better off than she was? My intuition is that she is not. And my indifference here is not just based on a view from the outside. While it makes no sense to ask whether I should care, as a nonexistent potential person, about being given a happy life, here I can put myself in Kate's shoes (who is an actual person) and feel unmoved by the choices given.

My indifference is also grounded in another, more important consideration: that a decision not to have children is not selfish, at least when it comes to consideration of the interests or happiness of those children that might have been, even if the prospective parents are totally indifferent about the choice. (Whether it is selfish relative to their community, let alone their own parents who may wish to be grandparents, is another matter.) This strikes me as a clear cut test case that counts against any theory that dictates otherwise. And that seems true of theories that count happiness untethered from actual people (now and in the future) as a virtue that trumps other considerations.

On Narveson's view the lives of others who exist can be made better as a result of someone being born, but his or hers cannot. For if it could be "then with whom, or with what, are we comparing his new state of bliss? Is the child, perhaps, happier than he used to be before he was born? . . . The child cannot be happ*ier* as a result of being born, since we would then have a relative term lacking one relatum" (Narveson 1967, 67). But notice, these considerations make no difference as long as we assume (as this literature does) that there *are* future generations. Narveson can be right that we do no good by bringing these future generations into existence, even if their lives are good. But that does not undermine our ability to assert that these generations that actually exist in the future are made worse off (or better off) than their lives would have been, but for our actions, even if their identities would have been different as well but for those actions.

But what if our actions now cause human extinction in the future? Then there are no future generations made worse off. Now Narveson's argument engages. There is no one who is made better or worse off by our actions. Could this be right? For it seems to sanction the following absurdity: *if there are future generations and we damage the planet, we do them a wrong. But if we damage it enough so there are no future generations, then we do not.*

But *is* this an absurdity? It certainly seems like it is when we focus on the degree of damage we may wreak on the planet and its consequences for those who may follow us, even if in doing so, the harm to those who may follow us is an unintended consequence of our actions. Now you might think intending such outcomes could make no difference or maybe even make this worse. But under *some* circumstances that is not the case and, in fact, we can undermine the absurdity. Suppose we knew that an action x would cause future generations to have a significantly lower quality of life than they would have had but for our action. Suppose we knew it would be significantly lower than ours is now, low enough to be a life worth not living. Say a life of perpetual hunger. On the other hand, suppose we knew that another action, y, would bring about total extinction of our species. And suppose those were the only two choices we had open to us. Now doing more damage seems more reasonable than doing less because it seems to do less harm.

But absurd or not, *can* we actually assert this on Narveson's view? For it seems equivalent to saying, given the choice between the outcomes caused by x and those caused by y, not being born (that is extinction) is preferable to being born (and living a life worth not living). But why doesn't "better not to have been born" succumb to the same considerations as "better to have been born"? If for it to be "better to have been born" requires a comparison that is not available to us, why wouldn't the same apply to "worse to have been born"?

What this demonstrates is that something has gone wrong here and to right it we need a more enriched way to think about lives led and not led than Narveson's conception allows. Here I follow McMahan (2009). Central to such an enrichment is the idea that even if a life cannot be better or worse in a comparative sense, it can be good or bad noncomparatively. So perhaps the life I lead today is better than the life I led last year. But the life I lead today is not better than the life I would have led if I had not been born. Nonetheless it is a good life. Still just this distinction is not enough. Both of these notions are person oriented, one comparatively and one noncomparatively. But what about a life not led that would have been bad had it been led? That life can't be worse than the life not led, for there is no life "not led" with which to compare. But it also can't be a bad life for someone because there is no one for which it is a life. So if it is bad it has to be impersonally bad. So now a life for a person can be comparatively good or bad, or it can be so noncomparatively for that person, or that life can be good or bad without being so for any person.

Now, if there are future generations, in damaging the planet, we can say we risk leaving them with bad lives, and in doing so, do them a wrong. But if we damage it enough so there are no future generations, then we can still say we caused a bad outcome. The first of these assertions can be understood in personal or impersonal terms. But the second can only be understood in impersonal terms. Have we thereby undermined the absurdity? If we have, this solution extracts a price. It runs contrary to what drew us to Narveson in the first place; namely, the intuition that making people happy is what matters. But at first glance there would seem to be an asymmetry that might salvage this intuition.

The following seems plausible to assert: if we were to produce future generations, we would have an obligation to ensure that they did not have lives worth not living. On the other hand, this does not seem plausible: if we could ensure that future generations that we might have had lives worth living, we would have an obligation to produce them.

This is a variant of this symmetry:

1. That a person would have a life "worth not living" ... provides a moral reason not to cause that person to exist.
2. That a person would have a life worth living does not, on its own, provide a moral reason to cause that person to exist.

(McMahan 2009, 49)

But McMahan argues this difference is illusory because it relies on an ad hoc distinction absent an argument to support it; namely, that being caused to exist can be a harm, but not a benefit. And moreover, if such an argument were available, it would support the implausible view that we should never have children since avoiding the possibility that a person would to have a life "worth not living" will always trump allowing for the possibility that a person would have a life worth living (McMahan 2009, 53–54).

But are these the only alternatives? McMahan argues that by relying on the moral difference between not causing harm and causing benefit, we can arrive at a third alternative. Namely:

> the reason not to cause harm by causing a miserable person to exist is stronger, perhaps considerably stronger, than the reason to bestow an equivalent benefit by causing a happy person to exist.
>
> (McMahan 2009, 57)

And what is important for our purposes is that that weak reason be weak enough to be easily overridden, as when a couple, knowing they would likely give a child a good life, nonetheless decide not to have that child merely as a matter of personal preference.

For now, let's not worry about what grounds the distinction between harming and not benefitting. For now we seem to have the best of both worlds with an account that allows that a good can be impersonal without embracing the consequence that we are under an obligation to maximize that good.

III

So far so good, but unfortunately, this leads us back to the absurdity albeit by a different route.

What would override the impersonal good that avoiding extinction and producing these future generations would yield? The most straightforward, if unlikely way to arrive at such a state, would be if, by chance, members of a whole generation happened to all share individual preferences against reproduction. Of course, deciding not to have children, when everyone else is also deciding not to have them as well, is a very different choice from deciding not to have them, assuming others will. Both existential and practical considerations may make it an implausible choice. But suppose it was the choice that everyone made. They looked at the existential void of being the last generation and embraced it. And perhaps technology might have advanced enough in their generation that robots could nurse them over death's door (see Bartz 2010). Existential and practical considerations notwithstanding, what matters more for our purposes is whether or not the force in favor of respect for reproductive freedom (not to have children) is weakened when it happens to be shared universally. Whether your interest in not having a child overrides the impersonal good that would result were you to have a child will be affected by the decisions of others to do likewise or not. The quality of life that your child will lead will be affected by the decisions of others. Your child might live in a very crowded world or a very lonely world. In this sense our decisions are never isolated one from the other. But if this affects the quality of life your child might have and hence your decision about whether or not to have a child, it does not affect your individual right to make such a decision. "What of a shared obligation to continue the human race?" someone might ask. "Surely that is what matters here." But what is basis for this obligation given our rejection that possible people have rights?

Suppose we don't all happen to decide not to produce future generations but, instead, act knowing that our actions may produce the same result. We allow our short-term interests to trump other considerations for one reason or another. Suppose we do so with the confidence that if things go badly, and our actions bring about extinction, we can manage the decline of our species to spare the last remnants of homo sapiens anything but a fun filled demise – a bacchanalian last blast. Then once again we may ask, whom have we wronged?

The answer might hinge on this: the road to the ruin of our species is unlikely to be selective, to just cause the demise of our species alone. Indeed on Hansen's scenario, we run the risk of causing the destruction of all life on Earth. What then of other species? Do we not wrong them by our actions? As with future generations, we can ask this with rights in mind (see for example Regan 1983), or approach the question side-stepping rights talk (see for example Singer 1975). I think both approaches face the same serious obstacle; unless you think nonhuman animals possess phenomenal consciousness, I don't think they have moral standing. This is not a position that helps me for it sets a high bar on grounding the moral standing of nonhuman animals. Were I to think animals had moral standing, it would be open to me to argue that even if we have no obligations to future generations of our own species were we to destroy the planet, we have obligations to existent members of other species.

I make phenomenal consciousness central to these considerations for the following reason: I treat a condition on moral standing as having interests (of the right kind), and I take a condition of having interests (of the right kind) to possess phenomenal consciousness. I assume stones lack interests and I assume that, to the extent that bacteria have interests, they are merely functional interests, like the interest in propagating. So what are the right kind of interests? Intuitively the idea is that how you are treated matters to you in a way that has no meaning when it comes to a stone or a bacterium. Feinberg (1980, 165) attempts to characterize the relative sense of interest as conative:

> mere things have no conative life; neither conscious wishes, desires, and hopes; nor urges and impulses; nor unconscious drives, aim, goals; or latent tendencies, directions of growth, and natural fulfillments.

But on that view bacteria have the relevant interests and so does every other living thing! Instead I think the relevant contrast is that even if a bacterium has interests, it does not care when those interests are not met. Who or what then does care? The traditional answer is that it is only those things with phenomenal consciousness and thus those for which it is something like to be those things.

But why not make feeling pain or pleasure the crucial feature? If we do then the question becomes what counts as "feeling." In particular, why should nociception (the capacity to sense noxious stimuli) without awareness count as feeling pain any more than it should for an unconscious human who nonetheless responds to being jabbed by a needle? If it doesn't then the pain that matters is the pain you can feel. But you can't feel such pain without awareness. Once you allow that some animals react to pain stimuli (by nociception) without awareness but that is not enough to give them moral standing on this account, the issue again becomes which animals, if any, have the capacity for phenomenal consciousness.

This is a far from fully satisfactory condition on moral standing as is since temporally unconscious people lose moral standing. Although we can fix that by extending the account to things with the capacity for phenomenal consciousness even if that capacity can't be expressed under all circumstances. What is trickier is when something has the potential to develop that capacity even though it does not have it now, as is the case, perhaps, with newborns or even fetuses. That puts us back in a position of according potential people standing, which is a position we previously rejected. To deal with these issues we need some distinctions that I will make use of first in a different context.

IV

Let us assume for the purposes of argument that no animals other than humans have a capacity for phenomenal consciousness and ask, if that is so, and we have no obligations to future generations of our own species were we to destroy the planet, would we do anything wrong in doing so? Look at pictures of large swaths of the Amazon that have been deforested. Look at the devastation, and can one deny that a wrong has been done? But whom or what has been wronged? Yes, the native people. And yes all of humanity that is surrounded by less of nature's variety and beauty. But these are moral judgments that revolve around us, around our needs. But what about nature itself and all of the clearly nonconsciousness species our actions are driving to extinction? Can we make sense of whether or not this is wrong without reference to what is good for ourselves?

You might think the prospects dim if phenomenal consciousness plays the role I have asserted. If we are skeptical about animal consciousness, then surely we ought to be so about plant life let alone "Nature" itself. But beginning with his seminal essay, "Should Trees Have Standing?," Stone (1996) took a different approach based on the history of the extension of rights to the previously deemed rightless. Those extensions were not based, he argues, on some universal property of rights bearers previously unrecognized in some class of the rightless, but rather on a much more gerrymandered approach. Historically,

Nonpersons were fitted into the Persons framework by overlooking distinctions that might have been drawn between them and Persons. Simply by the legal fiction of denominating the Nonperson a "person," it was thereby brought under the same principles as were applied to any "other" person.

(Stone 1987, 22)

In this sense, the extension of "personhood" to infants, the insane and most saliently corporations, provides a template for a further extension to not just trees, but nature as a whole.

"Rights" in Stone's sense are a matter of standing. To have such standing, a party must be adversely affected. Moreover, judgment would have to be for the benefit of that party. Thus even if such a party could not speak for itself, the damage and repair must be relative to its own interests. And there is the rub, as Stone admits: "Indeed, in what sense can we label as 'damages' alteration in the environment that no human even finds objectionable?" (Stone 1987, 10).

He offers two approaches to sidestep this problem that are not anthropocentric. One is to treat all life or nature as intrinsically good, while the other is based on transcendental considerations. The problem with the appeal to the former is that it fails to answer a skeptical stance. Suppose you say saving this river is an intrinsic good and you opponent asks why? What will you answer? An intrinsic good is not a good in virtue of its instrumental value. So your answer can't rely on that. The river must be good in itself. So far so good. But the problem that arises is one of the justification for that claim in the skeptic's eyes. The most plausible way to do that is by relying on the assumption that what is identified as "good" is self-justifying or needs no other grounding. The most obvious case is goodness itself. An appeal to its intrinsic goodness invites consent that this is where a chain of reductive moral justification might reasonably end. Yet the more we stray from claims about the goodness of goodness to other cases, the less obvious such an appeal becomes. Moreover, if respect for nature is an intrinsic good, then isn't respect for all its constituent parts – its cockroaches, its viruses, its maggots? The urge to ask "Why?" persists. (For the most ambitious attempt to develop such an approach see Callicott 2014.)

Alternatively, transcendentalism invites the idea that we efface the boundaries of the self not only with each other but also with nature. Our moral considerations will then attach to a wider subject of concern. Bracketing the question of whether or not this is not a variant of anthropocentrism, suppose we allow that we are, in some sense, one with nature. But to the extent that moral discourse is about conflict *between* moral agents and its resolution, we can't all be one with each other and with all of nature.

Suppose you are one with cockroaches. (Some grand Samsorian thought experiment.) Presumably your behavior toward them will now be constrained. But constrained by what? Not their interests, since if they had interests we would not be in the quandary we find ourselves in. It is the interests that arise from the fusion of you and them in to one unitary subject. The intuition here is that even if the cockroaches don't care about their well-being, the fusion of you and them does. I think this is a profound and subtle conception that recognizes the psychological reality of self-interest but attempts to transform it and thereby harness it. If we seek our own self-preservation, then we will seek the preservation of other species as well so long as we allow ourselves to fuse with them. And yet this still has the air of a world without risk of the kind of conflicts that give rise to moral discourse, a world in which it is us versus cockroaches.

Indeed, what got us going down this road in the first place was a conflict much closer to home: the idea that we might be willing to risk the continuance of our own species in the long-run because of considerations of short term self-interest. And it is in *that* context that we need to ask about the virtue of transcendentalism. Now perhaps with transcendentalism in hand we would be less prone to act on our narrow short-term interests. But that puts the cart before the horse. With our short-term interests fixed, it is those interests that will govern our attitude toward nature. What we sought was an account that would ground a moral stance in favor of preserving other species, even if we have no commitment to preserving our own species. It is hard to conceive of a version of transcendentalism that would do that. Instead, the fusion with future generations of our own species would seem a more plausible mental exercise before taking on fusion with the cockroaches of the world.

But even if we do take on an ambitious program of fusion, it is important to realize how profoundly conservative such a strategy for engendering respect for nature is. For what it does is to widen the scope of our desire for self-preservation of ourselves to nature *as it is now*. And that flies in the face of the most profound feature of nature; namely, its capacity for change, which cedes no privileged status to the manifestation of life as it happens to be now.

What would it be to take this feature and make it central to our thinking? It would be to take seriously the vast tableau of possibility that natural selection is able to exploit and amplify. We can stand in awe of this magnificent process and yet it is not a process that privileges any outcome over others. Still if we destroy all life, we end that process. Should we think of doing so as nothing more than an act of vandalism that offends us simply in virtue of its wantonness, one that reflects badly on our character by our own lights? But this is just anthropocentrism again. Whom or what do we wrong by such actions? Now it is here that the temptation of environmentalists is to do whatever needs to be done intellectually to make the answer "Nature!" (See for example, Universal Declaration of the Rights of Mother Earth 2010.) But philosophers writing on the environment have not been very obliging in giving a respectable basis for this answer. (For an overview see Jamieson 2008, 145–162.)

But what if instead of trying to give nature (or its current nonhuman constituents) moral standing, we focus on nature's *potential* to evolve *new* species that are aware, sentient beings that can express their interests like we can. For when we destroy all other life, we preclude the opportunity for beings like these to evolve. And it is *that* act of interference that I will argue is what makes our destruction of nature wrong.

Suppose we as a species left the scene, leaving the rest of life on Earth largely intact. There is of course no guarantee that a new species of aware, sentient beings with interests would arise. (That is beings for which there would be something like it is to be them.) But let's allow that there is at least a nonzero probability that they could. Call these Homo novus. They may not look or think like us, but what they will share in common with us is whatever we take to be the set of properties that make for moral standing that, for example, we would accord visitors from outer space. Members of Homo novus are potential people, like future generations of Homo sapiens, except that the probability of their existing is unknown, just assumed to be greater than zero. The question at hand is whether the possibility of their existence should constrain our actions now any more than the possibility of the existence of future generations of our own species. It better, since we concluded that the latter places us under at best weak constraints that are overrideable. Is there a difference between the two? Neither have rights to the extent that they are possible people, nor is "good" or "bad," good or bad for them. In both cases we have to construe "good" or "bad" in impersonal terms. Assume that in the case of future generations of our own species, not producing them denies good lives and is thus an act of not benefiting as opposed to harming (which would occur if we produced future generations with bad lives). Suppose the same is true of other species. Our actions deny good lives not bad lives. Then is this too not an act of not benefiting and hence easily overridden? Something is wrong here. Even if the lives we deny in both cases are good lives, surely there is a moral difference and it is one we can't capture without relying on something more than just the outcomes rather than *how* they are arrived at.

The idea that not causing harm is more important than causing benefit, all other things being equal, seems like an uncontroversial moral claim and one that needs no grounding. But to stop there is to leave us impotent in distinguishing between what concerns us here: contrasting cases of not causing benefits. Even if we reject the idea of harming and benefiting as needing no grounding to distinguish between them, the conventional basis for doing so is to appeal to the distinction between positive and negative duties. In the most straightforward cases, like limiting a person's freedom of movement against their will, when we harm someone, we do so by an action. That action interferes with the other and, in doing so, violates a duty not to so act – a negative duty. On the other hand, when we fail to benefit someone, like not helping to save a drowning man, we do so by inaction and, in doing so, violate a duty to act – a positive duty. That negative duties are stronger than positive duties gets its most salient support from the idea that each of us has a right to freedoms consistent with the same freedoms for others and that it is the demand for consistency that generates negative duties, that is, duties of noninterference. On the other hand, positive duties get no traction in this context. But one way they do get traction is as an extension of the idea that we should do to others as we would have them do to us. But in that case, the asymmetry of numbers dilutes its force as compared to negative duties, even to the point of rendering them as supererogatory. That is to say, since any one of us can interfere with another, the duty of non-noninterference has to apply to all of us with equal force. On the other hand, to the extent that there is a duty to help another, it applies to all of us collectively, except insofar as it has been explicitly assigned to one of us (as in the case of a lifeguard on a beach as compared to all the sunbathers).

Feinberg (1984, 170–171) would have none of this, arguing instead that there is no difference in the "force" of negative over positive duties except in so far as we have made an assignment of special responsibilities to some when it comes to positive duties, which only then diminish the force of the obligation for the rest of us. When you and only you can easily save a drowning child in a swimming pool and fail to do so, Feinberg would have us treat this as much more than a failure to engage in a supererogatory act (Feinberg 1984, 130–131). But we may agree with the latter without agreeing with the former. That is to say, we may treat positive duties as (at least sometimes) obligatory without necessarily treating them as having the degree of obligation as negative duties. All the more so, you might say, when it comes to causing a "happy person to exist." Whatever our treatment of other positive duties, like saving a drowning child, if causing a happy person to exist is a duty, it is one we need to treat as able to be overridden, assuming we want to preserve the right of reproductive freedom, at least when it comes to the freedom not to reproduce.

But on the other hand, when it comes to our preventing happy persons from existing in the case of other species, even as a matter of restricting the probability of this occurring, *it is not positive duties but duties of non-interference that engage.* And these create much higher expectations that are not so easy to override as a function of our preferences. In the case of our own species we can prevent future generations by choosing not to act (to have children). While in the case of other species, we can only prevent future generations by acting to prevent others from acting. Here then we have two contrasting cases of not causing benefits, one toward possible members of our own species and one toward possible members of another (possible) species that carry with differing degrees of moral force. Notice the underlying principle here that is doing the work is not just that negative duties are stronger than positive duties but that the failure to cause a happy person to exist cross-cuts this distinction. As such, how strong is its claim on us is a function of which of these duties is implicated.

My notion of a duty of noninterference toward possible beings is of course all pervasive. It creates a polymorphous prescription for us to tread lightly everywhere in the world, to look with awe at what might be. That will strike some as a tortured way to avoid a much more natural prescription: to simply *look* with awe at what *is*. In one sense this second is a much better prescription since what it holds out is the prospect of truly inspiring us with a thick and real sense of awe. On the other hand, my prescription is bloodless and brittle, a thin construct on an intellectual argument. My argument can't compete with the emotions that a walk among the giant redwoods in Muir Woods evokes. And while I can try to marry the two (by thinking about nature's possibilities as I walk in the redwood groves) that seems like a prescription to undermine both the clarity of my thinking about what could be, and the quality of my feeling about what is. Perhaps it would be better to keep the two apart. But perhaps we have been too quick to dismiss the possibility of invoking a sense of awe about what is possible rather than about what is.

V

Thoreau writes, in *The Maine Woods*:

> It is difficult to conceive of a region uninhabited by man. We habitually presume his presence and influence everywhere. . . . Nature was here something savage and awful, though beautiful. I looked with awe at the ground I trod on, to see what the Powers had made there, the form and fashion and material of their work. This was that Earth of which we have heard, made out of Chaos and Old Night. Here was no man's garden, but the unhandseled globe. It was not lawn, nor pasture, nor mead, nor woodland, nor lea, nor arable, nor waste-land. . . . Man was not to be associated with it. It was Matter, vast, terrific, – not his Mother Earth that we have heard of, not for him to tread on, or be buried in, – no, it were being too familiar even to let his bones lie there, – the home, this, of Necessity and Fate. There was there felt the presence of a force not bound to be kind to man. It was a place for heathenism and superstitious rites, – to be inhabited by men nearer of kin to the rocks and to wild animals than we.
>
> (Thoreau 1988, 94–95)

I have been arguing for two possibilities of unknown probability. But both Hansen's scenario and the evolution of Homo novus are more than logical possibilities. They are consistent with our physical laws and initial conditions reflected in the state of the world as it is now. Of course other possibilities are much more likely. Perhaps our destructive behavior will be less extreme that Hansen fears and, even if it jeopardizes our species, it will not jeopardize all life. Were we to pass from the scene leaving other species alive, it is unlikely that the Maine Woods would be reconstituted as they were before man. But, as Alan Weisman points out in *The World Without Us* (Weisman 2007), even in Manhattan change would occur quickly: within 36 hours the New York City subway system would be underwater (Weisman 2007, 25). Within 20 years, the girders supporting Lexington Avenue would collapse turning it into a river (Weisman 2007, 26). Within 200 years: "Arriving black locust and autumn olive shrubs [would] fix nitrogen, allowing sunflowers, bluestem, and white snakeroot to move in along with apple" (Weisman 2007, 28).

But such scenarios should not blind us to the moral consequences that attach to the more extreme possibilities we have been considering that countenance the elimination of all life on earth. Whether we choose to heed to those consequences as individuals matters less than whether we can do so collectively. And it is to that problem that I now turn. However before doing so I need to take up three loose ends to the argument so far.

Someone might object as follows: "You talk of negative and positive duties. But to every duty there is a right. If we have duties toward possible beings then they have rights. But you denied them rights earlier." The only way to answer this question is deny the first step. That is to deny that rights and duties go hand in hand. The most straightforward way to see this is in cases of supererogatory duties. If I do have a duty to save a drowning victim, that victim has no right to my exercise of such a duty. (For more see Donnely 1982 and more generally Lyons 1970.) Notice that there is a parallel here to our earlier discussion of the impersonal good.

Assume that both newborns and fetuses of our species lack the property of personhood (because they lack consciousness) but have the capacity to develop it over time. If duties are not treated as correlative in with rights, I can still have duties governing my treatment of both fetuses and newborns without according them rights. But must they not be the same as duties I have toward future generations that I have argued are (only) positive duties and hence limited? But of course there is a difference. In the case of newborns and fetuses duties of noninterference engage as well. (Although in the case of the latter, this is not to say they cannot be overridden given the special circumstances of a fetus's dependence in utero. See Thomson 1971.) But what about duties of noninterference toward future generations? Where these exist they hold a force no different to the force of those duties towards newborns or other species. So, for example, were some of us to decide on voluntary extinction and others not, then our actions would be constrained by negative duties of noninterference that would be harder to override than the positive duties we have been considering when unanimity reigns.

I said my notion of a duty of noninterference toward possible beings is of course all pervasive. It creates a polymorphous prescription for us to tread lightly everywhere in the world. But how lightly? Any action of ours in the world changes its course. Some possibilities are closed off and others are opened. In this sense, all of our actions interfere with some world lines and enable others. They deny the possibility of sentient beings coming to exist on some world lines and make it possible for others to exist along others. But if noninterference trumps action, and the two go hand in hand, then it would seem we should tread as lightly as possible wherever we act in the world.

References

Bartz, Daniel 2010. "Toyota Sees Robotic Nurses in Your Lonely Final Years," *Wired*, January 19, www.wired.com/gadgetlab/2010/01/toyota-sees-robotic-nurses-in-your-lonely-final-years/, accessed January 18th, 2012.

Callicott, J. Baird 2014. *Thinking Like a Planet*, Oxford: Oxford University Press.

Donnely, Jack 1982. "How are Rights and Duties Correlative," *The Journal of Value Inquiry*, 16 (4): 287–294.

Fehige, Christoph 1998. "A Pareto Principle for Possible People," in Christoph Fehige and Ulla Wessels (eds.), *Preferences*, Berlin: Walter de Gruyter, 508–543.

Feinberg, Joel 1980. *Rights, Justice, and the Bounds of Liberty: Essays in Social Philosophy*, Princeton: Princeton University.

Feinberg, Joel 1984. *Harm to Others*, New York: Oxford.

Hawken, Paul 1994. *The Ecology of Commerce*, New York: Harper.

Jamieson, Dale 2008. *Ethics and the Environment*, Cambridge: Cambridge University Press.

Lyons, David 1970. "The Correlativity of Rights and Duties," *Nous*, 4: 45–55.

McMahan, Jefferson 2009. "Asymmetries in the Morality of Causing People to Exist," in Melinda Roberts and David Wasserman (eds.), *Harming Future Persons, Ethics, Genetics and the Nonidentity Problem, International Library of Ethics, Law and the New Medicine*, Berlin: Springer, 45–68.

Narveson, Jan 1967. "Utilitarianism and New Generations," *Mind*, 76 (301): 62–72.

Parfit, Derek 1984. *Reasons and Persons*, Oxford: Oxford University Press.

Regan, Tom 1983. *The Case for Animal Rights*, Berkeley: University of California Press.

Singer, Peter 1975. *Animal Liberation*, New York: New York Review of Books.

Stone, Christopher 1987. *Earth and Other Ethics*, New York: Harper and Row.

Stone, Christopher 1996. *Should Trees Have Standing?* New York: Oxford University Press.

Thomson, Judith Jarvis 1971. "A Defense of Abortion," *Philosophy and Public Affairs* 1: 47–66.

Thoreau, Henry David 1988. *The Maine Woods*, London: Penguin.

Universal Declaration of the Rights of Mother Earth 2010, http://climateandcapitalism.com/?p=2268, accessed January 31st, 2012.

Weisman, Alan 2007. *The World Without Us*, New York: St. Martin's Press.

4

THE THREE TROPES OF CLIMATE CHANGE

I

Whatever our obligations to others (and nature) may be, if we have any, what then? How should we divide up those obligations? In examining this question I am going to challenge three prevailing tropes. The first trope is that the problem of climate was caused by the Developed World and so it is up to the Developed World to solve the problem. The second trope is that clean energy is available, it is just a question of whether we are willing to pay the higher costs for using it and who should pay those costs. The final trope is that the interests of the Developing World and the Developed World are aligned in that we all lose in the face of climate change. I take issue with all three of these claims. It is not that they are false, but rather that they are incomplete. Taken at face value, and so put to use in discourse about climate change, they foster the illusion (in the Developing World) that growth and avoiding climate change are not in conflict. Instead, it is a matter of added cost, which can and should be paid for by the Developed World. In the Developed World, these tropes foster the illusion that avoiding climate change trumps development for everyone, but especially the poor. And since the poor are overwhelmingly concentrated in the Developing World, so too, in the end, avoiding climate change will trump development in the Developing World as a whole. So when the parties meet, one side assumes the other can and should pay while the other side assumes that, if they don't pay, there is not too much to lose, since it is in nobody's interest to follow business as usual, that is, to fuel growth while continuing to emit carbon.

Let us examine these considerations one at a time.

II

You and I share a well. We both thought it would never run dry. I used a lot more water than you did. Now we discover it is going dry. We need to proceed with care, limiting our use. The well reaches an underground lake that is fed by a spring. We have been drawing water at a rate that outpaces how fast the spring replenishes it. So be it. But who will get how much? I make the following proposal: let each have an equal share going forward, be it per household or per person. You object. You have used a lot less than me over the years. You think that should be part of the accounting. "Yes," you say, "let there be a fair share, but looking back as well as forward." But until now, neither of us knew the well might run dry. The supply seemed inexhaustible. So my using more than you did not seem to matter to either of us. The idea of a "fair share" makes no sense if what is to be shared is inexhaustible. But it wasn't inexhaustible, we just thought it was.

Should ignorance matter here? Does it make any difference? Ignorance is said to be no excuse in the law. But in the case of the law there is something to be known that it is my responsibility to learn about. Here we had an unknown. Even though neither of us realized it, I got an unfair advantage. Unfair or just lucky? Suppose we say it *was* luck. I happened to be in the right place at the right time to make use of more of the water than you. Does that luck play a role in the moral equation of who should get what going forward? Does it insulate us from the charge of unfairness, so that a case of unequal distribution is not unfair if it is arrived at by the mere luck of one party over another?

We like to think we should only be held to account for what we have control over. And how can I have control over that which I am ignorant of and could not even have known if I had wanted to? True, when you and I do the same thing and, by pure luck, your act has consequences that mine does not, things get more complicated. A child runs in front of your car not mine. We were both driving with reasonable vigilance. For better or worse, it was your bad luck that you killed the child. In such cases luck does not function to insulate you from any moral responsibility at all, even though your act and mine were identical. But cases like these are about moral culpability while in the case of the well we are not worrying about moral culpability but a fair distribution. Still, even here, far from insulating us from the charge of unfairness, luck egalitarians would say that luck can be part and parcel of it.

Luck egalitarianism is an extension of an egalitarian conception of equity. As my colleague Larry Temkin (2011, 51) puts it, "[a]mong equally deserving people, it is bad, because [it is] unfair, for some to be worse off than others through no fault or choice of their own. But among unequally deserving people it isn't bad, because [it is] not unfair, for someone less deserving to be worse off than someone more deserving, even if the former is worse off through no fault or choice of his own." The role of distributive justice on this conception is to balance these inequalities and by extension to limit the effects of luck. But this is far from the end of the story, for how we may feel about fairness and luck may turn out to be a function of the details:

- You and I share a well which we assumed had an unlimited supply of water. I used more than you. I like children, you don't. I used the water to grow crops to feed my children. Now a fair share that looks backward as well as forward will not yield enough for my family. Our plans were made in good faith and upending them will have dire consequences. That seems unfair.

- You and I share a well which we assumed had an unlimited supply of water. I used more than you. I had a swimming pool and fountains, you did not. A fair share looking forward alone will leave my pool empty. So be it. A fair share looking backwards will hurt even more than that. But as long as I am guaranteed some supply of water to satisfy my basic needs that does not seem unfair. I enjoyed my pool. Now it's your turn.

What is the difference? In both cases, if I had known the truth about the water supply I might have acted differently – had fewer children or not built the pool. But what is done is done. Yet in one case the plans I laid can be aborted. The pool just stands empty. The children are another matter.

The details also matter when we look at things from your end:

- You and I share a well which we assumed had an unlimited supply of water. You used less than me. Your interests lay elsewhere. You could have used more. You simply chose not to. Here including a backward looking claim seems gratuitous.
- You and I share a well which we assumed had an unlimited supply of water. You used less than me. You wanted to do more, but you needed to spend your money on medicine for a sick spouse. Now you have the funds to invest and you want to make up for lost time.

I benefited from your situation in both cases. But of course here the difference is that in the second case you had a plan, albeit a frustrated plan. In the first you simply squandered an opportunity.

If you share my intuitions in these cases, they show how mere luck does not make a difference. When it comes to a fair distribution, the details matter. But enough with the well. If the details matter we need to trade the metaphor for that for which it stands, the atmosphere and who has put how much CO_2 and the like into it. I'll be the Developed World, the North, the West, the First World. You be the Developing World, the South, the Third World. (We could instead follow Chakravarty et al. (2009), and divide the world between individual high emitters, wherever they are located and everyone else based on national income distribution data. There is much to recommend proceeding this way looking forward, but it is hard to do so for past emissions because of the gross granularity of historical data.)

You say, "Look you have had your turn, now it's mine." But why is that the case? Perhaps simply because, in understanding our obligations toward each other, a prima facie principle of an equal share is all we need, at least when it comes to goods not subject to private ownership. But if that is plausible when it comes to we who live now, what about those who will be or could be, in the future? The future presents two distinct accounting problems. In the first instance we don't know how many people there will be in the future and how many there are will be a function of what we do now. As we have seen, if we are really selfish, there will be none. Let's assume there will be at least some. We don't need to know how many there will be in the future. But whatever the numbers, most of us would allow that we owe it to them to do no more harm and do our best to undo the harm we have done. But there is also a short range issue of the future. Suppose we say, an equal share for all alive now even if we respect a "do no harm" principle for future generations. We know that world population is rising and even if we pledged to stop its growth, it would take decades to stabilize. Fine, let's take that into account and include the hoped for steady state population a few decades from now (say 2050) as the basis for an equal share in the here and now. That will give us time to set our house in order and rein in population growth beyond that date. And if some choose not to do so, let them not be rewarded with an equal share as their populations continue to grow.

But an equal share of what? (At least for the purposes of argument, let us assume that there is something still to be shared, contra Hansen's (2008) concern, that we are already past a safe limit.) If we only look forward the story goes like this: for better or worse, unknowingly (until late in the last century) we humans put out more greenhouse gasses than the world could absorb. Without a steady state between output and absorption, these gasses have moved from about 275 ppm of CO_2 to 400 ppm. If the consensus is right then anything above 450 ppm can have serious consequences, which gives us a budget of 50 ppm of CO_2 which is equivalent to 106.5 gigatons of carbon (GtC). But because some of what we have put into the atmosphere comes out every year (absorbed by photosynthesis and air-sea gas exchange) we have a budget of (at most) twice that figure, 213 GtC or 781 gigatons (Gt) of CO_2, given the current levels of atmospheric concentration of CO_2, after which we will have to limit our annual output to be no more than what (net) ocean and terrestrial uptake affords us. CO_2 uptake is, in part, a function of its concentration in the atmosphere. (Here I am grateful to Ben Kravitz.) At 450 ppm, the ongoing allowance to maintain an equilibrium state would be roughly 5GtC per annum. However, once there, if the level proved to be too high, to reduce back by say 50 ppm would take 100 years even if we stopped emitting CO_2 altogether (see Solomon et al. 2009). Although that is not to say the effects of a 450 ppm world would be quickly eliminated. Any CO_2 removed from the atmosphere into the ocean has a very long cycle (on the order of 40,000 years) before all its warming and acidification effects are eliminated through absorption into the sea bed (Archer 2009, 108–113). So a steady state of 450 ppm will still produce ongoing cumulative warming of the oceans for a long time. That said, whatever the final budget is does not really matter for our purposes. For what we are pondering is how to share it, whatever it is.

So here is what might seem a plausible proposal: we will take the net greenhouse gas budget we have available to us and divide it equally between the expected population of 2050 which will be about 9 billion people. A better formulation would take account of the fact that we start at 7 billion with a population that will grow each year until it stabilizes at 9 billion. To accommodate that we should add the expected population for each year and divide the budget by that figure which would give everyone an equal per annum share. But once we do that going forward, why not do so going backward and account for those in the past who deserved an equal share per capita per annum that they did not get?

There are two issues here, one of which brings us back to the question of whether the past should count or not. The other adds the issue that if it does, for whom does it count? Thinking of historical claims of nation states masks this second issue. It invites the illusion of a fixed cast of characters acting through time. You (the Developing World) say I (the Developed World) used more than my fair share. Now it is your turn. Even if we ignore population movement (for example from the Developing World to the Developed World), what is the basis for your claim? Is it that your ancestors did not get their fair share which is now your due? Or is it that you have a direct claim? You say, "Look, it doesn't matter. You have had your industrial revolution, now it is my turn." This is a claim that the past matters. It is your turn you say because your ancestors did not have their chance and thus pass on the benefits to you. But it is also your turn because you have not had a turn. Either way you ground it, what matters you say is that you live in a society with none of the benefits of an industrial revolution and I do. So, you say, if we are to have an equal share, don't just look forward look back as well. But that is far from the end of the story.

"Not so fast!" I say. As we saw in the case of the well, when it comes to a fair distribution, whether history matters or not may depend on the details. And depending on the details, it may not be your turn above and beyond moving ahead on a forward looking equal basis. But the case for that is hard to make here. I'd have to argue that your ancestors squandered an opportunity that was theirs for the taking and that my ancestors acted reasonably and set us on a course that is hard to undo. Both of these fly in the face of plausibility. The counter claim to this need not be that, but for the oppression and exploitation of the Developed World, the Developing World would have thrived. Nor need it be that we can easily wind back the path of development of the Developed World to undo all of its dependencies on energy. A middle ground on both will suffice. It is hard to dispute the fact that the development path of the Developing World has been significantly shaped at least to some degree for the worse by the actions of the Developed World. And it is even harder to dispute that the development path taken by the Developed World to date is somehow unrevisable going forward.

If anything, the details of history will help not hinder a general argument that if we look forward, we should look backward as well when it comes to counting out a fair share on an equal basis. But can we do that, and if so, how much difference does it actually make? Peter Singer (Singer 2004) has given an eloquent defense of an equal division going forward, but he thinks that looking back is unreasonable if only because it is too complicated to calculate. I don't say it is simple, but it is not that complex either. However to do so we need to decide on one thorny issue. We want to calculate how much the Developed World has strayed above its fair share historically. To do that we need to know the ratio of the Developed World's population to the Developing World's. Should we do that based on the world's population now or look at the data for projected population over some reasonable time horizon? There is even a third alternative, which is I look to the world's historic population. For now, let's do it using the current population ratios and then see how much the alternative approaches make a difference. For comparative purposes, we should start with Singer's forward looking only approach. The world population today is (roughly) 7 billion people. At this point we need to become a bit specific (if arbitrary) about what counts as part of the Developed World. I am going to include the following:

Europe
North America
Japan
Australia
New Zealand

That gives us a population total for the Developed World of roughly 1.227 billion in a world of 6.705 billion or 18.17% of the population (Population Reference Bureau 2008). Current annual CO_2 output for the world (using data from the United Nations Statistics Division 2008) is 29.888 Gt of CO_2, of which the Developed World's share is 11.764 or 39.4%.

As we saw earlier if we wish to stabilize at levels no higher than 450 ppm, then the world's carbon budget is 50 ppm. Were we to divide that budget up on equal per capita terms looking only forward, what would the Developed World's share be and how many years would that last at current output levels? Again, allowing for terrestrial and ocean uptake of atmospheric CO_2, 50 ppm is equivalent to 213 GtC or 781 Gt of CO_2. That still leaves a stark difference between current Developed World consumption and its budget if we make allocations on an equal per capita basis which is 142 Gt of CO_2 – a budget that will be exhausted in a mere twelve years or so based on current annual output of 11.764 Gt of CO_2.

So far we have been looking at an accounting based on the Developed World's share of the current world's population. What happens if we look at things in terms of the projected population of 2050?

In 2050 the World population is expected to be 9 billion of which the Developed World will only comprise 13.5% (United States Census Bureau 2012). So if we use the 2050 projected population as our basis for an equal share, the Developed World's carbon budget is even smaller: 105 Gt of CO_2, which would last about nine years at current level of output. (Notice then that while natural gas produces only 50% of the carbon output of coal fired plants, replacing coal with it is only of limited value in buying much time. See United States Energy Information Agency 2012.)

But now what if we look at this looking backwards as well as forwards? Up until now we have been working with a carbon budget based on where we are today (400 ppm) and how high we can safely go (450 ppm). But looking back we want to start at pre-industrial level (275 ppm). The historical budget is 450–275 ppm. That gives us 175 ppm, of which the Developed World's share (based on current population) is 18.17% or 31.8 ppm. That translates into an historical budget of roughly 497 Gt of CO_2 (again incorporating an allowance for ocean and terrestrial reabsorption). But, relying on data from 1850–2002, the Developed World's output was 663 billion tons (even ignoring data prior to 1850, which is hard to get for many countries) and thus already far over its fair share (Baumert et al. 2005, 122).

Finally, looking at these calculations based on the world's historic population ratio has little material effect on the outcome. If we take 1850 as our starting point again, Europe and North America constituted 23.9% of the world's population, and together with the rest of what we have been calling the Developed World, 36%. So the Developed World's total share of the carbon budget would be 711 Gt of CO_2, leaving only four years left given its current rate of output.

So, only looking forward, the Developed World has a modest budget of carbon, while looking in both directions it is already in negative balance (unless we base the calculation on historic population ratios). Considerations of fairness would thus dictate that the Developed World has been, and is, living on borrowed CO_2 as it were, and that the balance of who owes whom what tilts radically in favor of the Developing World. Nonetheless, I want to argue this is a temporally parochial view in the following sense: looking backward from a different vantage point will yield radically different results. We only have to go forward a short period in time and the balance of who will have done what between the Developed World and the Developing World will most likely be very different, based on some plausible assumptions. (What follows is based on the research of Wheeler and Ummel 2007.)

Wheeler and Ummel use the historic output of the Developed World and the Developing World and project them into the future using the Intergovernmental Panel on Climate Change's (IPCC 2007) business as usual scenario (BAU). Using BAU for this exercise is an act of pessimism in that it assumes no effective transformation of the global economy to make it more ecologically friendly, while it also assumes continued economic integration and a slowing of population growth but at the same time, a priority on economic growth. In doing so BAU assumes a central role for fossil fuels in realizing these goals. Later on I will defend why I think BAU is the right scenario that we should focus on, but for now, let's examine the consequences when cumulative emissions from the Developed World and Developing World are separated in this scenario for the rest of this century. Then the Developing World's contribution to atmospheric CO_2 looks like this:

Year	ppm
1980	300
2000	325
2040	375
2060	450
2080	550

This table is worth keeping in mind as one reflects on a report of India's UN Ambassador, speaking on April 20, 2007. He "told the developed nations that the main responsibility for taking action to lessen the threat of climate change rests with them ... while efforts to impose greenhouse gas commitments on developing nations would 'simply adversely impact' their prospects of growth" (Wheeler and Ummel 2007, 1, quoting Press Trust of India/Factiva, April 20, 2007). For these projections show that bracketing all of the Developed World's output, past, present, and future, the Developing World is on track to produce enough output in and of itself to create a problem as shown in Figures 4.1 and 4.2.

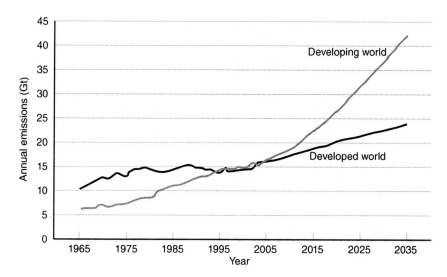

FIGURE 4.1 Annual CO_2 Emissions from the Developing and Developed Worlds (Wheeler and Ummel 2007)

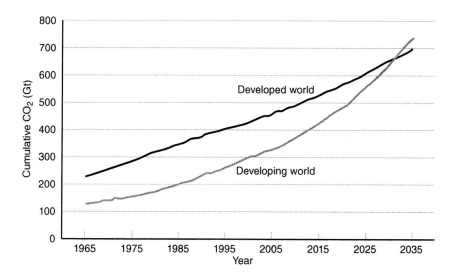

FIGURE 4.2 Cumulative Atmospheric CO_2 from the Developing and Developed Worlds (Wheeler and Ummel 2007)

Thus irrespective of what the Developed World were to do, including offsetting its historic output, you still don't sidestep the problem. In this sense, the problem is not just a Developed World problem. Indeed looking back from just 2030, the cumulative historic output of the Developing World overtakes that of the Developed World.

III

But notwithstanding the question of who did what, when, and who owes what to whom, what if we resolved (collectively) to move forward without increasing our carbon footprint and even aiming to reduce it? Much of climate policy frames this question as a matter of will and a matter of cost. Climate policy assumes that demand *can* be met. The only question is whether we choose to do so with clean or dirty sources. How realistic is that assumption?

In the set of projections for contrasting emissions scenarios, developed in the year 2000 by the IPCC (2000), the business as usual model (BAU) was conceived of as a model of rapid economic growth in which *coal* plays a primary role, and was only later combined with other fossil fuels into BAU. Data sets for BAU use the different models (AIM, MESSAGE, and MINICAM). For simplicity here I rely on just one of them (MESSAGE), (although any of them would do because they are close enough to each other for our purposes) which yields key figures for the state of the world in 2100:

> Population – 7.056 billion. (The model assumes population peaks in 2050 at 9 billion and then declines.)
> GDP – 535 trillion US $. (In 1990 prices, using purchasing power parity.)
> Primary Energy – 2325 EJ. (Where 1 exajoule (EJ) = 1018 joules.)
> Cumulative CO_2 – 2112 GtC. (Where 1 gigaton (GtC) of carbon = 3.664 Gt of CO_2 and 2.13 GtC = 1 ppm by volume of atmosphere of CO_2.)
> Temperature rise – 2.4°–6.4°C. (The likely range of change over the century with 4°C as the best estimate. See IPCC 2007.)

These models assume an average worldwide growth rate of about 3% and so the crucial question is whether such a rate of growth could be sustained without incurring the risk of the projected associated temperature rise. Another model (BTN) quantifies the cost of avoiding this risk, absent widespread substitution of clean technology for fossil fuels, with a lower rate of growth but instead emphasizing environmental sustainability:

> Population – 7.056 billion.
> GDP – 318.8 trillion US $.
> Primary Energy – 755 EJ.
> Cumulative CO_2 – 837.4 GtC.
> Temperature rise – 1.1–2.9°C.

These models date from 2000 and based on actual primary energy data to date (available from the International Energy Agency (IEA) 2012), BAU, if anything, understates primary energy demand so that BAU gives a conservative projection of the effects of business as usual:

BAU (2010, projected)   IEA (2010, actual)
511 EJ   533 EJ

Be that as it may, the contrast in coal consumption between BAU and BTN is itself extreme:

BAU (2100)   BTN (2100)
1062 EJ   22 EJ

And all of this applies with even more force in Asia where we expect the largest economic growth this century:

BAU (2100)   BTN (2100)
218.2 US$ 103.1 US$

driven by a regional contrast in coal use:

BAU (2100)   BTN (2100)
386 EJ   10 EJ

The contrast between BAU and BTN offers a stark tradeoff between GDP growth and the projected climatic consequences. But the family of scenarios offers a third alternative (BOT) which appears to allow us to have our cake and eat it as well in 2100:

Population – 7.056 billion.
GDP – 535 trillion US $.
Primary Energy – 2021 EJ.
Cumulative CO_2 – 1061 GtC.
Temperature rise – 1.4–3.8 °C.

So what does BOT assume to overcome the climate change cost of BAU and the forgone economic growth of BTN? A contrast between the sources of primary energy for the models tells the story in 2100 as seen in Table 4.1.

Here, the category of "other renewables" is doing most of the work, 61.3% to be exact, although BOT does not treat the category of other renewables as significant until late in the century. Earlier in the century, oil and gas do most of the work, as seen in Table 4.2.

TABLE 4.1 Primary Energy in 2100 by Scenario (IPCC 2000)

Source	Energy produced (EJ)		
	BAU	BTN	BOT
Coal	1062	22	25
Oil	56	46	77
Gas	118	215	196
Nuclear	432	41	114
Biomass	376	235	370
Other Renewables	281	197	1239
TOTAL	2325	755	2021

TABLE 4.2 Primary Energy in 2030 by Scenario (IPCC 2000)

Source	Energy produced (EJ)	
	BAU	BOT
Coal	261	180
Oil	210	223
Gas	207	231
Nuclear	41	40
Biomass	95	104
Other renewables	57	73
TOTAL	870	850

How realistic is this scenario? We can approach this question from two different directions. First, by asking about the model's long-term prospects when it comes to the renewables and nuclear portfolios. Second, by asking how this model compares to the actual record to date.

In BOT, by the end of the century, renewables and nuclear power constitute 67% of the energy portfolio. Be that as it may, the most comprehensive examination of how to get from here to there for a nuclear-renewables scenario looks at the issue through the long term lens of energy that is *fully* carbon free. That said, we can put its detail to good use here.

In his book *Sustainable Energy – Without the Hot Air*, David MacKay (2009, 237–238) argues that non-solar renewables (including wind energy) provide a maximum total of exploitable energy of 9×10^3 GWh or about 24 kWh per person per day (assuming a midcentury population of 9 billion instead of the population of 6 billion he uses in his calculations). That is far below the current European level of 125 kWh per person per day as well as the level MacKay defends as realizable with plausible conservation measure; namely, 80 kWh per person per day. There are only two ways to make up this difference without reliance on fossil fuels: solar power and nuclear power. And in fact it is possible to do so with either if we ignore costs. Erecting mirrors to concentrate solar power for steam generation is much less efficient than photovoltaic systems, but the former suffers from no constraints on the supply of raw materials unlike the latter. Ignoring transmission losses, MacKay calculates that we would need to build two solar arrays each 1,000 km by 1,000 km in a place like the Sahara Desert (MacKay 2009, 178). (His calculation is to produce 125 kWh per person per day for a population of 6 billion. But the same array would support 9 billion were we to reach his conservation goal of around 80 kWh per person per day.) However transmission by high-voltage direct current adds another 15% to demand to compensate for transmission losses (MacKay 2009, 178–179).

That said, a set of solar arrays of this size (roughly four times the area of France) would not only be an enormous undertaking, but would not be able to satisfy base load needs, the minimum needed to satisfy demand at any time, including at night, absent the development of massive storage technology. Absent that, the only non-carbon source of energy available to us in the quantities needed is nuclear energy. Current nuclear (fission power) generation is constrained by two factors: the type of plant and the source of the material. Most nuclear plants are light water (once through) reactors which use much more fuel than fast breeder reactors. Based on current technology, a one-gigawatt light water reactor uses 162 tons of uranium a year. Mackay posits total land uranium reserves in the ground to be 4.7 million tons with 22 million more tons extractable from phosphate deposits. Here (with the assistance of Joe Reilly) I use the lower Organization for Economic Co-operation and Development (OECD) estimate a total of 13 million tons from all land sources (OECD 2010). That would provide 223 years of nuclear energy at current rates of energy production of 2,938,304 GWh. But divided between the current population of 7 billion, that would only take care of about 1.18 kWh per person per day, far short of the 80 kWh per person per day and even less (.9 kWh per person per day) for the expected world population of 9 billion in 2050. Were we to rely on such reactors for all of our prospective power needs, assuming again 80 kWh per person per day, and a population of 9 billion, it would last only 2.5 years! On the other hand, fast breeder reactors are 60 times more efficient than light water reactors, so the same energy would sustain us for a much more comfortable window of 150 years. The figures used here for uranium extractable from phosphates are based on imposed limits for the costs of extraction and could thus be stretched by increasing the costs. However doing so does not increase the totals available by an interesting order of magnitude. That said, uranium exists in much greater quantities in the sea than in the ground – at a ratio of roughly 1000:1 compared to land sources. However the rate of extraction is limited by how slowly oceans circulate (once every 1600 years). Assuming 4.5 billion tons of total ocean deposits, that would provide 2,800,000 tons of extractable uranium per year. Assuming (following MacKay) a 10% extraction rate (ignoring costs) would thus yield 280,000 tons or 1772 GWh when used in a once through reactors, and this would only produce about 5 KWh per person per day for a population of 9 billion. On the other hand, the deployment of fast breeder reactors together with use of uranium extracted from the sea would ensure an adequate supply of power for about 50,000 years for a population of 9 billion at 80 kWh per person per day.

So, ignoring cost, if we chose to do so, we could *eventually* satisfy our prospective energy needs (as estimated by MacKay). Moreover, these sources of energy would allow us to satisfy those needs even if we have underestimated demand, it is just a matter of deploying more solar mirrors or increasing the intensity with which uranium is derived from sea water for more fast breeder reactors. Indeed that is clearly going to be the case if we take into account some ancillary considerations that MacKay himself recognizes ought to be added to total energy consumption above and beyond his figure of 80 kWh per person per day. Of these the largest is the embedded energy in imported goods. If we look at the per capita energy use of a North American or European country, which largely exports services or information but imports manufactured goods, the net energy balance of the later over the former ought to be taken into account. MacKay sets the figure for the embedded energy of imports and their transportation at 60 kWh per person per day. Do we need to do the same for a global energy calculation? We surely do, since, if we aspire to have a world in which all people live at the standard of living of European using conservation, then those people will be consuming manufactured goods at the same rate as European even if they are manufactured domestically. The other factor that needs to be taken into account is the conversion of energy from one kind to another for end use consumption as well as transmission losses of electricity. MacKay calculates the first at 22% and the second at 1%. However the figure for transmission losses over long distances were we to use solar arrays in desert areas for most of our energy would be around 15% by his own admission.

So while we have been using 80 kWh per person per day as our guide, the real figure we need to use adds another 60 kWh per person per day for embedded energy inflated by another 37% to take into account conversion and transmission, which gives us a total of 192 kWh per person per day or almost two and a half times the figure we started with! With that in mind the size of the solar arrays and the lifetime of nuclear supplies would need to be adjusted accordingly. That said, of course, in the long run, the prospects for better conservation, more efficient energy production and lower transmission losses make all of this an arcane exercise. But it is not an exercise without merit, for what it shows is that, if we were so disposed, we *could* supply all of our energy needs to satisfy a world of 9 billion people living at a Developed World standard of living, without any carbon output. Of course that is a largely nuclear world with its own set of risks. But, even on a worst-case scenario, those are risks for limited populations and limited areas of the world, unlike the risks posed by carbon output on the worst case scenarios we have been considering.

But irrespective of which figures we use, 80 kWh per person per day or 192 kWh per person per day, notwithstanding the merits of examining long term carbon free energy supplies, it is the short term that provides the greater challenge. For it is not just a matter of getting from where we are now to 80 kWh per person per day, but doing so on a timetable compatible with demands to do so at an acceptable *rate* of economic growth. At first glance, that may not seem a challenge if we look at the current distribution of energy consumption worldwide (ignoring embedded energy, transportation and conversion) using 2008 figures for kWh per person per day (IEA 2010):

USA	239
EU-27	112
China	51
Africa	21
India	17
World	58

Then if we take MacKay's figure of 80 kWh per person per day as our goal (bracketing embedded energy et al.), conservation by the United States and Europe would seem an obvious place to "free up" and redistribute existing energy instead of producing more. But, since Europe and the United States only constitute 11% of the world population, that redistribution is of limited value. Thus a reduction in U.S. per capita from 239 kWh per person per day to 80 kWh per person per day redistributed to the rest of the World, excluding Europe, results in a gain of only 7 kWh per person per day. The same exercise for European consumption produces an additional gain of 2 kWh per person per day. That may help, but it does not take into account the energy demand given the fact that the population will grow by another 2 billion in the next 30 years. Nor does it take into account that our current energy production is itself largely carbon based and thus in need of transformation irrespective of the demands of energy growth.

All of this suggests that the logistical challenge of building the necessary infrastructure to produce clean energy in the volume needed needs to be considered taking into account the timetable by which it would need to be in place to limit carbon output. Whatever timetable one selects for such a program, it extracts a price in terms of the diversion of resources and, obviously, the more aggressive the timetable the greater that price. But irrespective of cost, there are technical limits on just how quickly one could put in place the kind of massive solar and nuclear infrastructure MacKay advocates.

As Biello points out (Biello 2011), the United States would need to build 1,000 one-gigawatt nuclear reactors by 2050 to supply one-quarter of its *current* energy mix. That would be extraordinarily ambitious in and of itself but even more so given recent history. As Davis (2011) shows, of the 197 reactors on order in 1974, only half were ever constructed. Indeed no new construction of nuclear plants took place between 1977 and 2013. Moreover the U.S. Department of Energy lists only seven plants scheduled to come on line between now and 2023 (Department of Energy 2013).

What if those limits are not compatible with the point at which we exhaust our carbon budget? What if we use up our carbon budget and lack adequate supplies of clean energy to replace fossil fuel? Then if we chose to adhere to the strictures of that budget there will be a cost of foregone economic growth associated with the limits on the availability of adequate energy supplies. Short-term models hold out the hope that we can avoid this choice by moving to using less coal in favor of more natural gas and more nonfossil fuels between now and 2030. But, because the window for action here is so small, these models are highly sensitive to delays in implementation of any change from business as usual. As such, we need to take into account deviations between their assumptions of energy use and the actual record to date. Nothing makes this clearer that the IEA 2009 model for stabilization at 450 ppm of CO_2.

On the face of it, the IEA (2009, 191) offers a strikingly pain-free pathway toward a 2030 energy portfolio compatible with stabilization at 450 ppm as compared to a business as usual scenario which would put the world at 660–790 ppm by the end of the next century. "Business as usual" in the IEA's "Reference Scenario" assumes the continuation of policies already in place (IEA 2009, 173) which makes it somewhat more climate friendly than the IPCC's business as usual scenario (BAU). Be that as it may, the Reference Scenario assumes a 40% increase in energy demand (IEA 2009, 73) and a compound annual growth rate worldwide of 3.1% per annum, with China assumed to grow at 5.9% and India at 6.3% (IEA 2009, 63). The model assumes a changing energy mix, but one in which fossil fuels still dominate as seen in Table 4.3 with renewables and nuclear constituting roughly 20% of the total in 2030 and an anticipated change in energy related CO_2 emissions from 28 Gt to 40 Gt by 2030 (IEA 2009, 111).

In contrast, in the 450 scenario, energy related CO_2 emissions peak in 2020 at 30.9 Gt by limiting energy growth to 20% and increasing the proportion of zero carbon energy sources to 32% (IEA 2009, 195). The percentage contrast between the two scenarios highlights the shift away from carbon based energy as shown in Table 4.4.

Crucial to this plan are of course the nonfossil fuel capacity additions required as seen in Table 4.5, as well as the use of carbon capture and sequestration for 20% of added coal capacity.

TABLE 4.3 The IEA Reference Scenario (IEA 2009, 74)

Source	Energy produced (EJ)	
EJ	2007	2030
Coal	133	205
Oil	171	210
Gas	105	149
Nuclear	30	40
Hydro	11	17
Biomass	137	67
Other renewables	3	15
Total	503	703

TABLE 4.4 IEA 450 Pathway Compared to the Reference Scenario (IEA 2009, 212)

Source	Difference in 2030 scenarios
Coal	−47%
Oil	−15%
Gas	−17%
Nuclear	+49%
Hydro	+21%
Biomass	+22%
Other renewables	+95%

TABLE 4.5 IEA 450 Pathway Additions to Capacity (IEA 2009, 234)

Source	Change in Capacity (GW) 2008–2030
Nuclear	378
Hydro	832
Biomass	237
Wind	1129
Solar	502
Geothermal	32
Tidal	9

How realistic is this? The model assumes a price in that GDP in 2030 would be between .9% and 1.6% lower than it would have been in the Reference Scenario, which seems modest against the assumption that the world economy will double in size between 2007 and 2030 (IEA 2009, 203). But even if that price is acceptable, a key to the model is what it assumes about the use of coal in electrical generation by China (IEA 2009, 352) and India (IEA 2009, 360) in 2030 as compared to the Reference Scenario; namely a drop of 48% and 54% respectively from figures of 82% and 76%, and equally dramatically in IEA 2013 (623, 627) of 42% for both China and India from already lowered figures of 65% and 68% for 2035 when compared to the Current Policy Scenario that is in itself stricter than the Reference Scenario. Not only does this seem ambitious but it bears no relationship to either county's current energy and climate plans (see Communist Party of China 2011; Sharma 2012).

That said, in the case of China, *some* modeling suggests that a low carbon growth pathway *is* possible given a number of assumptions about the longevity of current observed trends in falling rates of urbanization due to rising rural wealth, increasing natural gas availability, a decreasing emphasis on heavy industry in the economy, high rates of innovation on renewables, and increasing rates of energy efficiency (see

Kejun 2012). However, the prospects for realizing such a long-term model, even if it is possible, depend on government policy action which is currently not forthcoming given government statements that China's carbon will peak between 2030 and 2040 as reported in the Guardian (Watts 2009). (I discuss the significance of this in Chapter Eight.) Moreover, the global carbon budget implications for such a model are unclear since it assumes a shift of heavy industry from China to less developed countries.

Models with a longer timeline for a transition away from fossil fuels (to 2050) may seem more forgiving. But as we already saw, looking at the first decade of the IPCC scenarios themselves, in the light of the actual record, the deviation between assumptions and the "business as usual model" (BAU) reality looms large. Thus to continue business as usual is to follow a course even more extreme than BAU in the IPCC scenarios. If we do that we use up our carbon budget sooner rather than later and BTN with its lowered growth becomes our fate. The choice between economic growth and restraining carbon output then becomes unavoidable.

But what about the promise of BOT? BOT is more restrained in its ambitions for nuclear power and renewables than MacKay's totalistic program. Nonetheless, comparing BOT and the IEA's 450 scenario, the near-term demands of BOT are even greater than the IEA 450 scenario in the scale of biomass renewable deployment it assumes (see Table 4.6).

More recent modeling incorporates variables for temperature sensitivity and alternative social policies but do little to undermine this choice when it comes to the short run. Thus for example, Eom et al. model a 9 billion person world with a 500 trillion dollar economy by the end of this century (POP9/MDG+) that has a "highly managed environment" (Eom et al., 2012, 33). They then project a range of transient temperature rise (depending on the temperature sensitivity) from pre-industrial times between 1 °C and 3.5 °C and a steady state rise over pre-industrial times between 2 °C and 6 °C (Eom et al. 2012, 51). This is based on a social policy (SPA 4.5) that all but eliminates carbon based sources of energy by 2085, except for 20 EJ of natural gas, in an overall energy budget for electrical generation of 550 EJ.

TABLE 4.6 The IEA 450 Scenario and BOT (IEA 2009, IPCC 2000)

Source	Energy produced (EJ)	
EJ 2030	IEA	BOT
Coal	205	180
Oil	210	223
Gas	149	231
Nuclear	40	40
Hydro	17	–
Biomass	67	104
Other renewables	15	73
TOTAL	703	850

But here too the timetable to build to an essentially carbon free economy is crucial since in SPA 4.5 renewables and nuclear comprise 50% of the energy budget for electrical generation.

It is not then the case that we cannot have a carbon-free energy system if we are willing to pay the higher price for it. Rather the issue is that the time needed to deploy such a system at the needed scale runs the risk of outpacing our carbon budget, given the current rate at which we are using it up. The incompatibility between these two forces a stark choice between sustaining economic growth and restraining carbon output.

IV

The Developing World has the most to gain from economic growth, but the vulnerability of its population means they also have the most to lose from climate change. What if a choice has to be made between the two? But choice for whom? The Developing World is far from homogenous, as are the populations of the countries that make it up. As such, the answer to this question may vary across countries that make up the Developing World and across populations within those countries. For countries like the Maldives or Bangladesh, with extensive territory at sea level, the effects of climate change will surely trump increases in wealth through growth. And for countries with very uneven income distribution, the same may be true, at least for the majority of their populations, as the benefits of growth are concentrated in the hands of the wealthy. But let us suppose, even if it is only for the purposes of argument, that an enlightened leadership of a country is faced with the choice of promoting growth in income for its poor at the risk of precipitating climate change, or foregoing both. We need not assume the country has income parity. The rich may get richer through growth, but the focus of the cost-benefit analysis, as it were, brackets that. For now, let us also assume that *all* the poor stand to gain from such a program to promote growth, albeit at the risk of climate change. That is to say, for now, let's assume that there may be losers, but if there are, they are in the future not the present. Still this is not a cost-free choice for people now, especially poor people, to the extent that people in the future may include their offspring. If we think of this calculus in these terms, I want to argue that it is far from obvious that foregoing growth is the rational thing to do.

Notice that this is not Bjørn Lomborg's question about the cost of diminishing the risks of climate change in Lomborg (2001). Lomborg defends the view that the risks of significant climate change damages are low and considers whether the cost of diminishing that risk further would not produce greater social benefit if put to a different use, for example, eradicating disease. I am concerned with the question of what to do if economic growth is impossible (at the desired rate) without risking (significant) climate change and how *that* choice should be evaluated.

The Nullah Leh flows through Rawalpindi. In good times it is the source of nurturance. But in bad times it is the source of destruction. Mustafa Daanish (Daanish 2009) has argued that those who live by the banks of the Nullah Leh will experience climate change, not as an additional new source of destruction but as an exacerbation of existing sources of destruction: the river will flood more often and to higher levels. Consider yourself a householder located near the river. Assume you have no funds to spare for home improvement as things stand. But now suppose you are given the following choice: an increase in income with a greater chance of climate change or holding to the status quo. An increase in income will allow you to build a new home on stilts to better withstand the flooding. But climate change may increase the severity of that flooding. So the stilts will have to be higher and stronger than what would be needed without climate change. That said, you may still come out ahead. If enduring climate change is the only way to increase your household income to have funds to build in the first place, the key question will be what the *additional* cost will be of building to withstand the effects of climate change on the river. There is an underlying intuition behind the idea that you may still come out ahead here: that climate change poses less of a risk than other threats that poor people face in their lifetimes. If there is a choice of increasing income with a risk (or even the actuality) of climate change, that income gives you protection not only against the effects of climate change, but also against more salient risks as well. If that rings untrue to citizens of the Developed World, it maybe because they (or we) start with a relatively high household income. Additions to that income that risk climate change are relatively small relative to our total income, as well as relatively modest in their value to us. A $1,000 addition to the household income of a $100,000 a year household has a very different value than it does for a $1,000 a year household.

The edge that a poor household lives close to is marked by vulnerability to disequilibrating threats like disease, theft, and unemployment that loom much larger in daily life than climate change. In contemporary thinking, this vulnerability is *the* defining feature of poverty rather than low income per se (see for example Sen 1992). The poor face greater risk in their daily life than the wealthy, but they also lack the resources to withstand the effects of those risks when they are realized. Hence they are more vulnerable, be it to the hazards posed by chronic risk or the shocks posed by stochastic risk (see Wood 2003, 459), who argues that "the vast majority of poor people face chronic rather than stochastic insecurity" and what this implies for public policy). If we think of vulnerability as inversely related to resiliency to risk, mere exposure to risk is then not necessarily synonymous with poverty. Rather what also matters is the ability to recover from the hazard and shocks posed by risk. Wealth makes a difference here, but absent wealth, other social and cultural endowments can act as substitutes. Thus, in the extreme, Bollig (2006) describes pastoral societies in which the ethos of sharing wealth provide all members of the society with cushions against shocks except for those that affect the society as a whole.

But such collective arrangements are of course far from the norm. Both historical sources (for example Davies 2010, originally published in 1795) and contemporary surveys (for example Collins et al. 2009) testify to the tightrope on which most of the poor walk daily in managing finances and how easy it is to fall off it. Jane Pryer's study of household economies in the Dhaka slums (Pryer 2003) highlights how much difference small amounts to income makes. But it does so in a surprising way. Being better off does not insulate you from getting sick as seen in Table 4.7.

Instead, it insulates you from the extreme effects of illness that affect the poor in terms of days lost, especially when it comes to female headed households that are the poorest, as shown in Table 4.8.

Illness of the income earning member(s) of a household is the leading cause of the deterioration of the financial stability of households in Pryer's study (21.6%) followed closely by wage or earnings decrease (19.3%) and the inability to find work (18.1%). (See Pryer 2003, 164 taken from Table 144.) And of course by their nature, these second and third causes of the financial stability of households affect the poor more than others because of the nature of their employment. Where poor people have work, it is more likely to be irregular and subject to loss when days are missed due to illness.

However we define the poor, of all the risks they face, none poses greater consequences than the death of offspring that they expect to depend on in old age. The risk benefit analysis of the stilts assumes the effects of climate change are limited, limited to an additional rise in the river. But the value of avoiding the loss of one's immediate offspring may hold even if we treat the risks of climate as potentially unlimited in the long run, as we have been assuming up until now. For avoiding loss of my immediate offspring obviously trumps avoiding the risk of loss of future generations that are my direct descendants since their very existence depends on the existence of those immediate offspring. (Unless that loss is compensated by the survival of many more indirect descendants that bear a close genetic relationship to me.)

TABLE 4.7 Household Adult Illness and Wealth (Pryer 2003, 88 taken from Table 7.6)

Adult members reporting sick in last 14 days	Incidence in %	
	Above the poverty line	Below the poverty line
At least one	81.4	82.4
At least half of all adult members	76.3	74.3

TABLE 4.8 The Cost of Ill Health (Pryer 2003, 99 taken from Table 8.3)

	Self-employed	Casual skilled	Causal unskilled	Female headed households
Total workdays off per month due to illness	2.24	2.89	4.2	7.58

The idea that increasing risk is worthwhile, if the income gain outweighs it, underwrites many of our daily choices. When you drive to work, instead of staying home, your risk of death increases, but presumably the benefits of the income you gain (and could not gain staying at home) outweigh those risks. Indeed, in properly working job markets, all other things being equal, salary differentials between different jobs ought to reflect risk differentials. (See the earlier discussion of the statistical value of life). That said, actual market conditions often don't work properly and people are forced to assume increased risk without a commensurate gain in income. The most obvious case of this may seem to be the "urban penalty" that characterized the industrial revolution. But that said, the data on urban penalties over time offers a rich resource to better understand just where and how the poor face a *rational choice* between growth and risk in the real world.

Writing about filth, noise, and stench in early urban England, Emily Cockayne (2007) cites a saying that suggests that even if conditions were worse in urban areas as compared to the countryside, people accepted that as the cost to have a prospect of riches: "We will bear with the Stink, if it bring but the Chink" (Cockayne 2007, 242). But "stink" was the least of their problems. Child mortality in urban settings in England ran far ahead of rural settings, both before the Industrial Revolution as shown in Table 4.9, and well into it as well, as seen in Table 4.10, which makes it hard to see why such an urban penalty would be voluntarily embraced. Of course on the standard view of the causes of industrialization, the rural poor were forced off the land and out of villages by land enclosure. Although as Hobsbawm (1969, 120) points out, that was but a symptom of the underlying cause which was the "concentration and consolidation" of farming. Still whatever the cause, there was no element of choice. But that won't explain the above data for the period 1500–1800, unless aggregate mortality does not tell all of the story of urban versus rural settings.

One thing the data might mask is a difference in the data as a function of wealth that might form the basis for an explanation. If the better off do a lot better than the poor, even if they are small in number, moving to the city might be a rational "bet." That is, a bet that one might end up among the better off and that the improvement in one's life would make that bet worthwhile. But that can't be right if we can rely on the data provided by Clark and Cummins (2009) seen in Tables 4.11 and 4.12.

TABLE 4.9 Implied Life Expectancy at Birth 1500–1800 (Clark and Cummins 2009, 244 taken from Table 1)

Area	Life expectancy (years)
London	22.6
Town	34.8
Rural	40.5
Farm	42.8

TABLE 4.10 Expectation of Life at Birth (taken from Szreter and Mooney 1998, 88 Table 1)

	1851–1860	1861–1870	1871–1880	1881–1890	1891–1900
Bristol	39	40	42	46	47
Sheffield/Newcastle	36	35	37	40	42
Gateshead	35	35	38	41	43
Leeds	36	35	38	40	42
Bradford	37	36	38	42	44
Birmingham	37	37	39	42	42
Manchester	32	31	34	37	36
Liverpool	31	30	34	36	38
London	38	38	40	43	44
England and Wales	41	41	43	45	46

TABLE 4.11 Probability of Survival to Age 25 by Wealth and Location, 1500–1800 (Clark and Cummins 2009, 249 taken from Table 2)

Asset income	London	Town	Rural	Farm
£0–6	0.40	0.59	0.61	0.64
£6–13	0.47	0.66	0.70	0.73
£13–31	0.42	0.61	0.67	0.71
£31–	0.44	0.63	0.72	0.75

TABLE 4.12 Number of Surviving Offspring by Location and Asset Income, 1500–1800 (Clark and Cummins 2009, 249 taken from Table 3)

Asset income	London	Town	Rural	Farm
£0–6	1.10	1.78	2.05	2.36
£6–13	1.49	2.37	2.65	3.03
£13–31	1.56	2.46	2.83	3.58
£31–	2.03	3.51	3.60	3.97

These data seem to fly in the face of such logic of making any such bet. Being poor on a farm trumps all outcomes in London irrespective of the income differential! Moving to town, betting you will be better off, would not then be a rational choice if mortality is what matters. Nor would it be rational for the rich to remain in town. These considerations may suggest that something is askew with the data here which are drawn from wills. (Reliance on wills creates two potential sources of bias: those with income are more likely to make wills than those without, but more significantly, those with higher net fertility will be more likely to make wills than others, irrespective of income. If the latter is significant, the testimony of the wills will overrepresent (largely rural) poor families with higher net fertility, relative to the underlying population of rural poor families, as compared to (largely urban) richer families, relative to the underlying population of urban rich families. I am grateful to Alice Reid here.)

Whatever the differences between the urban rich and poor, both were affected by the elevated health risks of urban living, especially with regard to epidemics before the advent of public health works in the late 19th century and 20th century advances in health care (see for example Checkland 1964). The latter helped reverse the aggregate "urban penalty" not just because it heralded discoveries to combat disease but because medical services could be provided more efficiently in urban settings in comparison to rural settings (see Leon 2008).

But even in the modern era, the reversal of the urban penalty does not hold across all incomes. Thus the National Research Council (2003) offers a striking example in Table 4.13.

If we assume that the rural poor first move to urban slums here too, as in the historic cases, they are (on average) worse off than they were in rural settings. So, here too, to be rational a move from rural to urban settings for the poor, if an act of choice, has to be based on a bet that one will end up among the better off; that is, as counted among the "urban" as opposed to the "urban slum." Indeed, a number of studies of urban migration treat the idea of a move from rural to urban settings as just such a bet in formal terms (see for example Todaro 1971) or embrace it as an assumption (see for example Islam and Azad 2008).

Even if it is on an idealized basis, viewing rural-urban migration as a lottery in which only some win, but the probability of winning and the payoff make it worth taking the risk of losing, can be broadly generalized to other kinds of risk taking, including the one that concerns us here, risking climate change for the potential gain of wealth, especially for a poor person deciding on a strategy that maximizes the chances that his or her progeny will survive and successfully reproduce themselves. At least it can, given the constraint we have placed on the argument when it comes to the choice between growth and avoiding the risk of climate change.

TABLE 4.13 Intraurban Differences in Infant Mortality Rates Bangladesh per 1,000 Live Births in 1991 (National Research Council 2003, 285 Table 7-7)

	National	*Rural*	*Urban*	*Urban Slum*
Total	90	93	68	134
Male	98	97	70	123
Female	91	89	65	14

V

Risk taking makes rational sense as a function of the payoff except when one is a sure loser, that is, when the probability of the payoff is zero. In the discussion so far we have assumed that when it comes to climate there may be losers, but if there are, they are in the future not the present, and, at least in the extreme case of extinction, their standing has been questioned. But what if the choice of growth over climate change is worth it for some of the poor but not for all of them? What if there are sure losers now? From a moral point of view we face the classical issue of whether policy should be determined by its effects on the worst off or some assessment of overall or average benefit. A popular interpretation of John Rawls' work (Rawls 1971) is that when it comes to the allocation of economic benefits and burdens, the right policy is always that in which the worst off are in the best off of worst off positions. (Although Rawls himself defended the principle as a macro principle not properly applicable to "small-scale situations," Rawls 1974, 142.)

Of course, as we noted earlier, the idea that one would be better off with more wealth and climate change is a function of your local circumstances. If you live in Bangladesh your vulnerability to even small rises in sea level will likely make for a very different calculus than one if you live inland and above sea level. The discourse of climate change tends to focus on these the most vulnerable (the worst off) as opposed to considering the overall or average consequences. As Oxfam reports (Schuemer-Cross and Taylor 2009, 2):

> Each year, on average, almost 250 million people are affected by 'natural' disasters. In a typical year between 1998 and 2007, 98 per cent of them suffered from climate-related disasters such as droughts and floods rather than, for example, devastating but relatively rare events such as earthquakes . . . by 2015 this could grow by more than 50 per cent to an average of over 375 million affected by climate-related disasters each year.

And yet, these additional 125 million people represent only about 2% of the Developing World, far less than even what Paul Collier terms "The Bottom Billion" (Collier 2007). Still whether small in relative number or not, should the needs of these, the worst off, trump other considerations? Of course we might be able to sidestep this choice even if they are made worse off by assuming that we could provide them with offsetting benefits. The same kind of issue arises everyday in more prosaic settings. Suppose you and I live in a village. The majority of villagers want to build a road to get the harvest to market. The only problem is that the road will go through your fields. You will be worse off with the road, but the rest of us will be better off. Should we let the numbers decide? Do we have a right to proceed? Most philosophers would argue, even if rights are relevant at all here, we do have a right to proceed because we are not interfering with any of your basic freedoms, like speech, movement, and the like. Rather, we are interfering with economic well-being. But that is surely too easy. Here there is likely more at stake. Whole communities will be uprooted. Ways of life will be destroyed. People may even die. And yet, sometimes, even such extreme outcomes get put on a balance scale for consideration of offsetting compensation. Thus, in an inoculation program, we know some (even if very few) will be permanently disabled or even die in reaction to the shots they get, and we accept this as necessary for the greater good and offer compensation for the harm we do. But what if such offsetting compensation is not possible or what if we simply reject the notion that it can ever adequately right a balance scale against the kinds of losses we are considering? Should we nonetheless allow the needs of the many to outweigh those of the few?

Harking back to our earlier discussion of a lifeboat in space, in so-called "lifeboat" ethics cases, two competing intuitions often coexist when you try to think of a fair distribution of a resource. One is to say everyone should have an equal share. The other is to look to the consequences and try to maximize them. In a first course in philosophy, I ask my students to consider such a case where the only food resource is each other. The task is to determine a fair order of consumption. (Although Talmudic reasoning says the proper thing to do under such circumstances is to all starve to death instead. In the Palestinian Talmud, Terumot 8:12, there is discussion of a case in which a non-Jewish ruler goes to a town and asks them to turn over one Jew for death, or he will kill everyone. The sages argue for letting everyone die rather than picking out one individual.) Most of my students say, "draw straws" at first. But a few call for eating the oldest (who has less years to lose), or the heaviest first (who may give us enough food to last until we are, hopefully, saved). What is interesting is that as I differentiate the cast of characters, more and more of the students who had previously favored drawing straws come to favor looking to the consequences. Let José one of the characters, be terminally ill and some will change their minds. Increase the number of children that will be orphaned if we eat Mary over childless Joseph, and more students will change their minds. By the time we consider Maria, who has the cure for AIDS and world hunger in her head but no paper to write it down, nearly everyone is a consequentialist. Indeed, my own philosophical prejudice is to say that in lifeboat cases like this, except at the extremes, where we would all stand to lose because of the prospective contribution of the individual, everyone deserves an equal chance, be it by drawing straws or any other random selection mechanism.

What is going on here? It may look like a debate about justice or fairness. Is the just or fair decision one that looks to the consequences or one grounded in a procedure that ignores the consequences? A consequentialist may defend drawing straws as a right thing to do under such circumstances because in the long run it will produce the greatest good. But the deontologist may retort that that is not *why* it is the right thing to do. Here the deontologist and consequentialists are debating whether or not justice is grounded in welfare. (Here and throughout when I use the term "welfare" I mean considerations that maximize benefit or the good.) Perhaps my students and I are through and through consequentialists who think drawing straws usually produces the best outcome but not always. Hence our changeability.

But that is certainly not how I think. I think drawing straws is *always* the just and fair thing to do. However I also think that (sometimes) considerations of welfare override considerations of justice. Thus the question or not whether or not justice is grounded in welfare, but how to balance them as competing considerations, and nowhere is that more true than when it comes to social policy that involves large numbers of people.

In contrast Mary Robinson (as in for example, Robinson and Miller 2009) has championed viewing climate change exclusively through the lens of considerations of justice:

> Climate justice as a concept represents the confluence of different streams of concern with fairness and ethical relations as they relate to people's use of the world's finite carbon resources.

In doing so, she embraces the idea that:

> human rights analysis and advocacy have always paid particular attention to those who are on the margins of society as a result of poverty, powerlessness, or systemic discrimination.

Rejecting this view is not to belittle the hardship these 125 million may face but rather to raise a consideration that that they suffer hardship while others may benefit is not the end of the matter. Deciding on the basis of the worst off positions does not always trump other considerations. Does not or should not? That I and my students may allow welfare considerations to trump considerations of justice doesn't end the matter. Should we? Here I think we are on soft philosophical ground. We do well in balancing competing considerations of justice and in balancing competing considerations of welfare, but poorly in balancing the two against each other. To echo Thomas Nagel (Nagel 1979) in this sense we lack a genuinely universal moral theory. But although you and I may differ as to when considerations of welfare trump considerations of justice, I am willing to wager that all of us will do so at some point when the numbers grow large enough.

But even if that is the case, *are* the numbers large enough here? Historically the preconditions for economic development have favored income inequities as a means of capital accumulation. Be it by way of robber barons or despots, the accumulation of wealth deprives the poor of satisfaction of their immediate needs thereby, at least in theory, freeing capital for long term investment. An enlightened state can do the same thing by placing an emphasis of economic development over the current well-being of its poorest citizens. Economic development for whom though? A realistic accounting of statecraft is likely to show that the beneficiaries are the already privileged classes. But they are not the only beneficiaries, as forcefully argued by Gregory Clark (2007) in his history of world economic growth since 1800.

VI

We have been examining the idea that short-term interests may trump long-term interests for the poor on the assumption that a clean energy supply cannot reconcile these two interests, at least on a timetable dictated by the needs of the poor. But in doing so, we have been very much looking from the outside in, even when it comes to judgments of rational choice. What happens if we look at the same choices from the inside out? I want to discuss how the circumstances of poverty shape the perceived calculus of risk and in doing so, further tip the balance in favor of the short term, for better or worse.

References

Archer, David 2009. *The Long Thaw*, Princeton: Princeton University Press.

Baumert, Kevin A., Tim Herzog, and Jonathan Pershing 2005. *Navigating the Numbers: Greenhouse Gas Data and International Climate Policy*, Washington, DC: World Resources Institute, 122.

Biello, David 2011. "Green Energy's Big Challenge: The Daunting Task of Scaling Up," *Environment 360*, http://e360.yale.edu/feature/green_energys_big_challenge__the_daunting_task_of_scaling_up/2362/, accessed November 5th, 2012.

Bollig, Michael 2006. *Risk Management in a Hazardous Environment: A Comparative Study of Two Pastoral Societies*, New York: Springer.

Chakravarty, Shoibal, Ananth Chikkatur, Heleen de Coninck, Stephen Pacala, Robert Socolow and Massimo Tavoni 2009. "Sharing Global CO_2 Emission Reductions Among One Billion High Emitters," *PNAS*, 106: 11884–11888.

Checkland, Sydney 1964. *The Rise of Industrial Society in England 1815–1885*, New York: St. Martin's Press.

Clark, Gregory 2007. *A Farewell to Alms: A Brief Economic History of the World*, Princeton: Princeton University Press.

Clark, Gregory and Neil Cummins 2009. "Urbanization, Mortality, and Fertility in Malthusian England," *American Economic Review: Papers & Proceedings*, 99 (2): 242–247.

Cockayne, Emily 2007. *Hubbub: Filth, Noise and Stench in England 1600–1770*, New Haven: Yale University Press.

Collier, Paul 2007. *The Bottom Billion*, Oxford: Oxford University Press.

Collins, Daryl, Jonathan Morduch, Stuart Rutherford and Orlanda Ruthven 2009. *Portfolios of the Poor: How the World's Poor Live on $2 a Day*, Princeton: Princeton University Press.

Communist Party of China 2011. *The Twelfth Five-Year Plan for National Economic and Social Development of the People's Republic of China*, Chicago: MoraQuest LLC.

Daanish, Mustafa 2009. "Hydro-Hazardscapes of Pakistan: Redefining Adaptation and Resilience to Global Climate Change," Presentation at *The Fifth Magrann Conference*, Rutgers University, April 16th–17th, 2009.

Davies, David 2010. *The Case of the Labourers in Husbandry Stated and Considered*, Cambridge: Cambridge University Press.

Davis, Lucas 2011. "Prospects for U.S. Nuclear Power after Fukushima," *Energy Institute at Haas Working Paper Series*, http://ei.haas.berkeley.edu/pdf/working_papers/WP218.pdf, accessed August 15th, 2013.

Department of Energy 2013. http://energy.gov/ne/downloads/quarterly-nuclear-deploy ment-scorecard-july-2013, accessed December 1st, 2013.

Eom, Jiyong, Richard Moss, Jae Edmonds, Kate Calvin, Ben Bond-Lamberty, Leon Clarke, Jim Dooley, Son H. Kim, Robert Kopp, Page Kyle, Patrick Luckow, Pralit Patel, Allison

Thomson, Marshall Wise and Yuyu Zhou 2012. "Scenarios of Future Socio-economics, Energy, Land Use and Radiative Forcing" in R. G. Watts (ed.), *Engineering Response to Global Climate Change: Planning a Research and Development Agenda*, Boca Raton: CRC Press.

Hansen, James 2008. "Climate Threat to the Planet: Implications for Energy Policy and Intergenerational Justice," Lecture given Dec. 17 at the American Geophysical Union, San Francisco. (Slides posted at www.columbia.edu/~jeh1/presentations.shtml, accessed June 4th, 2011.)

Hobsbawm, Eric 1969. *Industry and Empire*, London: Penguin.

IEA 2009. *World Energy Outlook*, Paris: IEA.

IEA 2010. *2010 Key Energy Statistics*, Paris: IEA.

IEA 2012. *World Energy Outlook*, Paris: IEA.

IEA 2013. *World Energy Outlook*, Paris: IEA.

IPCC 2000. Nebojsa Nakicenovic and Rob Swart (eds.), *Emissions Scenarios*, www.grida.no/publications/other/ipcc_sr/, accessed April 3rd, 2009.

IPCC 2007. *IPCC Fourth Assessment Report: Climate Change 2007*, www.ipcc.ch/publications_and_data/ar4/syr/en/main.html, accessed May 15th, 2011.

Islam Mazharul and Kazi Azad 2008. "Rural–Urban Migration and Child Survival in Urban Bangladesh: Are the Urban Migrants and Poor Disadvantaged?," *Journal of Biosocial Science*, 40: 83–96.

Kejun, Jiang 2012. "Secure Low-carbon Development in China," *Carbon Management*, 3 (4): 333–335.

Leon, David 2008. "Cities, Urbanization and Health," *International Journal of Epidemiology*, 37: 4–8.

Lomborg Bjørn, 2001. *The Skeptical Environmentalist*, Cambridge: Cambridge University Press.

MacKay, David 2009. *Sustainable Energy – Without the Hot Air*, Cambridge: UIT.

Nagel, Thomas 1979. *Mortal Questions*, Cambridge: Cambridge University Press.

National Research Council 2003. *Cities Transformed: Demographic Change and Its Implications in the Developing World*, Washington, DC: The National Academies Press

OECD Nuclear Energy Agency and the International Atomic Energy Agency 2010. *Uranium 2009: Resources, Production and Demand*, Paris: OECD.

Population Reference Bureau 2008. *2008 World Population Data Sheet*, www.prb.org/pdf08/08WPDS_Eng.pdf, accessed August 18th, 2012.

Pryer, Jane 2003. *Poverty and Vulnerability in the Dhaka Slums*, Aldershot: Ashgate.

Rawls, John 1971. *A Theory of Justice*, Cambridge: Harvard University Press.

Rawls, John 1974. "Some Reasons for the Maximin Criterion," *American Economic Review, Papers and Proceedings of the Eighty-sixth Annual Meeting of the American Economic Association*, 64 (2): 141–146.

Robinson, Mary and Alice Miller 2009. "Expanding Global Cooperation on Climate Justice," www.realizingrights.org/pdf/Climate_Justice_Robinson_Miller_Dec09.pdf, accessed June 17th, 2013.

Schuemer-Cross, Tanja and Ben Heaven Taylor 2009. *The Right to Survive*, Oxford: Oxfam International, www.oxfam.org/en/policy/right-to-survive-report, accessed August 6th, 2012.

Sen, Amartya 1992. *Inequality Reexamined*, Oxford: Clarendon.

Sharma, Shalini 2012. "Change Mitigation Strategies of India," Presentation at Department of Environmental Sciences, Rutgers University.

Singer, Peter 2004. *One World*, New Haven: Yale University Press.

Solomon, Susan, Gian-Kasper Plattner, Reto Knutti and Pierre Friedlingstein 2009. "Irreversible Climate Change Due To Carbon Dioxide Emissions," *PNAS*, 106: 1704–1709.

Szreter, Simon and Graham Mooney 1998. "Urbanization, Mortality, and the Standard of Living Debate: New Estimates of the Expectation of Life at Birth in Nineteenth-century British Cities," *The Economic History Review*, 51(1): 84–112.

Temkin, Larry 2011. "Justice, Equality, Fairness, Desert, Rights, Free Will, Responsibility, and Luck," in Carl Knight and Zofia Stemplowska (eds.), *Responsibility and Distributive Justice*, Oxford: Oxford University Press.

Todaro, Michael 1971. "Income Expectations, Rural-Urban Migrations and Employment in Africa" *International Labour Review*, 104 (5): 387–413.

United Nations Statistics Division 2008. *Millennium Development Goals Indicators*, http://mdgs.un.org/unsd/mdg/SeriesDetail.aspx?srid=749&crid, accessed August 18th, 2012.

United States Census Bureau 2012. *International Data Base*, www.census.gov/population/international/data/idb/informationGateway.php, accessed August 2nd, 2013.

United States Energy Information Agency 2012. *Frequently Asked Questions*, www.eia.gov/tools/faqs/faq.cfm?id=73&t=11, accessed August 20th, 2012.

Watts, Jonathan 2009. "China's Carbon Emissions will Peak between 2030 and 2040, Says Minister," *The Guardian*, December 6th, www.guardian.co.uk/environment/2009/dec/06/china-carbon-emissions-copenhagen-climate, accessed July 12th, 2012.

Wheeler, David and Kevin Ummel 2007. *Another Inconvenient Truth: A Carbon-Intensive South Faces Environmental Disaster, No Matter What the Developed World Does – Working Paper Number 134*, Washington DC: Center for Global Development, www.cgdev.org/content/publications/detail/14947, accessed March 13th, 2012.

Wood, Geoff 2003. "Staying Secure, Staying Poor: The Faustian Bargain," *World Development*, 31 (3): 455–471.

5

THE VIEW FROM THE INSIDE OF POVERTY

I

Living close "to the edge" the poor are exposed to greater risks than those who are better off, but, perhaps as important, they are also in less of a position to take risks as well. In this sense, the poor suffer from two traps; one is the more familiar trap. A household that lives on what it can grow and on the fuel (firewood) and water it can collect, faces many threats that can undermine its precarious welfare – not just illness or unemployment, but also natural disasters and (multigenerational) population growth without any commensurate increase in the land farmed. Even absent those threats, because the family economy produces no surplus, it is unable to take advantage of any opportunities that might generate savings through (say) trade which could then be put to use in ways that would increase that surplus (see Sachs 2005, 52–56). But even with (modest) savings, a family at or near the poverty level faces a different trap created by aversion to risk taking. Pryer (2003, 13) argues that as "households move closer to extreme poverty and destitution, they become very risk averse." Wood (2003, 455) argues that such risk aversion among the poor produces "dysfunctional time preference behavior." So it should not be surprising that, in poverty, time discounting of future income is overdiscounted relative to short-term self-interest, which creates a further basis for the poor to choose income now over avoiding climate change in the future.

II

Historically, the poor have been viewed through the lens of two very different models. One views them as rational agents, conforming to the axioms of rationality, but in circumstances in which the choices they face produce outcomes that seem irrational by our lights only because we are not exposed to the choices they face.

The other views them as less able than us as decision makers, so that what they choose in their circumstances is not what we would choose were we in the same circumstances. So on the first view the "overdiscounting" of future income is not irrational, but on the second it is. These two views carry with them very different explanatory stories as well. In the case of the former, circumstances are the relevant causal agents of difference. While in the case of the latter, it is a difference between agents that makes the difference.

But neither of these alternatives is satisfactory. The assumption of rationality fails in the circumstances that the poor face if for no other reason that it fails for all of us, because none of us are rational agents, at least when it comes to conformity to the axioms of rationality. On the other hand, when it comes to the circumstances under which the poor make decisions, empirical research suggests that were we placed in similar circumstances our choices would not be as different as we might think they would be.

Let us consider rationality first. It is tempting to think that if preferences are not set, it ought to be possible to treat any set of choices as rational. That is to say, for any set of choices there is *some* set of preferences that renders those choices rational. But that is not the case. For what the axioms of rationality require is consistency over choices, irrespective of agent preferences. Now when it comes to time preferences, the poor are more likely to display a greater degree of discounting as compared to the norm in two respects. One is a matter of the discount rate itself. Future income in the present is discounted to reflect the time value of money and with it comes to the degree of discounting, the poor discount more than the rest (Meier and Sprenger 2011).

But the poor are also more likely to discount in a different way as well. Suppose I offer you $100 now or $101 in a month. You decide which deal to take based on your discount rate applied to that 1 month delay. Now suppose I offer you the same deal, but for 1 year down the road. That is, I offer you $100 in a year or $101 in a year and a month. Here too there is a 1 month delay that merits discounting but it is a year in the future. Now if you are an exponential discounter your discount rate for this second 1 month delay will be the same as it is in the first case. That is to say, your discount rate (whatever it is) will be time invariant.

But it turns out that for us all (poor or not) that is not the case albeit in varying degrees. A choice between two points in the future matters less than a choice between now and a point in the future all other things being equal. And the more highly we value the present over the future, the more exaggerated that difference will be. Now when it comes to rationality, the problem is that such (hyperbolic) discounting invites an inconsistency of preferences over time. Consider again the choice of $100 a year from now versus $101 in a year and a month. That choice between two points in the future will, over time, become a choice between a point in the present and a point in the future. It will happen precisely in one year. But when that happens, if I discount hyperbolically, the discount rate I used to make my initial choice will no longer accurately reflect the discount rate over the choice in the present. That produces an inconsistency in which at one time I prefer one outcome (X − $100) over another (Y − $101 a month later), but at another I prefer Y over X, merely in virtue of the passage of time.

On the other hand, the alternative view of the poor as less able than us as decision makers fares not better. The idea that what they choose in their circumstances is not what we would choose were we in the same circumstances is hard to test empirically. But a clever use of time as a proxy for money suggests that when we are placed in similar circumstances our choices would not be as different as we might think they would be. Shah et al. (2012) argue that what counts as "similar circumstances" can be understood in the most general terms; namely, simply having less. Shah et al. recruited undergraduates to play the game Family Feud in which teams (families) compete to correctly identify the most popular answers people have given in surveys. Two families compete against each other in a contest to name the most popular responses to a survey question posed to 100 people; for example, "name a word that most people yell at their dogs" or "if a witch was not paying attention to where she was flying, name something she might crash into." The majority of respondent answers for the first of these is "no" and "a tree" for the second (see Family Feud & Friends 2012).

Participants in these studies (Princeton University undergraduates) were assigned time budgets that varied in their size as well as whether or not they could be traded across rounds. Moreover, for some, borrowing was interest free while for others it was not. The focus of the study was on the behavior of those playing with imposed scarcity but with the ability to borrow. As predicted, across different variants of these experiments, these subjects did indeed perform worse than others. Moreover, those operating under conditions of scarcity who borrowed did worse than those who did not, even if they too were operating under conditions of scarcity. And those borrowing with interest did even worse. Just why this might be is something we will take up later, but here the key point of this study, and others like it, is that randomly assigned experimental subjects are subject to spiraling debt cycles which end them in poverty traps.

This finding suggests that it is not a different psychological state that the poor bring to the economic market place that is the source of the trouble, but rather it is the circumstances in which they find themselves in the market place that is crucial. But moreover if the degree of their behavior under such circumstances deviates from the axioms of rationality, there is no reason to think ours would not do so as well. That is to say, there is no reason to assume the poor to be any more or less rational than other economic agents, all of whom deviate from the canons of rational agency in predictable ways depending on particular circumstances.

III

But this raises a more general issue. Assume that none of us is perfectly rational even under the best of circumstances − not just when it comes to hyperbolic discounting but in other basic ways as well. Thus while the axioms of rationality call for consistency with respect to the transitivity of preferences, few of us can maintain consistency under all circumstances beyond a limited range of choice. (That is, if you prefer A to B and B to C and C to D ... and M to N, will you consistently prefer A to N?) Then the interesting question is just how the circumstances of poverty make those deviations from perfect rationality worse. That may seem like a singularly gratuitous question since the answer may seem self-evident: having less! But in fact that is only part of the story, and in a way it is the least interesting part.

In a number of papers, Mullainathan and his associates (Mullainathan and Shafir 2009) and Bertrand et al. (2006) have developed a theory of "slack" as a property the poor lack but the rest of us have in varying degrees. "Slack" is defined as "the ease with which one can cut back consumption to satisfy an unexpected need" (Mullainathan and Shafir 2009, 129). In the first instance, aside from income level, slack is a substitute for savings, for example an extended family, a community chest, or even an overdraft line of credit at a bank. But lack of slack characterizes the lives of the poor in much more subtle ways which have far reaching implications for how decision making is made by them. Of these, one of the most important is the effect that lack of slack has on long-term planning. An obvious reason for this is that less slack means there is less with which to plan. But there is also another more surprising reason as well. Lack of slack means more attention has to be given to short-term decisions. The really rich are lucky to never need to cut back on consumption. And for the rest with slack, much of consumption is still above and beyond the bare necessities of life, so that for them as well cutting back is not a matter of great moment, be it at Starbucks or on Amazon. When you have enough slack, it means you don't need to keep track of all of your current expenditures (except the big ones) because you can proceed with the confidence that there is enough play in your budget that no cutbacks will be required to cover such expenditures. You don't need to worry that each debit card purchase may overdraw your bank account, and, consequently, you don't need to spend the time checking your balance before each purchase.

But all of this is not just a matter of convenience. If you assume that we operate on a cognitive budget as well as a financial budget, lack of financial slack has significant cognitive consequences as well. To speak of a cognitive budget is to assume that decision making is as much of a limited resource as money is. One way to think of that budget is simply in terms of time. The time it takes to make some decisions competes with the time that could be made making other decisions. Under those circumstances, immediate decisions will crowd out long-term decisions simply because the latter are less pressing. How to juggle your budget to have enough food for both today and tomorrow will crowd out the decision of how to set aside some of your budget next week for a long-term emergency fund. And when next week arrives, the same conflict or a close relative of it will happen again.

But another way to think of a cognitive budget is not in terms of time per se but in terms of the sheer number of such decisions you can make in one time period even if you were to have enough time to make each of them. Interviewed by Michael Lewis, Barack Obama indicated that he was familiar with research on decision making:

You'll see I wear only gray or blue suits, . . . I'm trying to pare down decisions. I don't want to make decisions about what I'm eating or wearing. Because I have too many other decisions to make. . . . You need to focus your decision-making energy. You need to routinize yourself. You can't be going through the day distracted by trivia.

(Lewis 2012)

Dean Spears set up both lab and field experiments to test hypotheses about poverty and cognitive control, which Spears characterizes as "the cognitive process, associated with working memory, that directs attention and inhibits automatic behaviors to pursue executive goals" (Spears 2011, 5).

In Spears' experimental setup, a choice between cooking oil and an empty tiffin (or alternatively a length of synthetic rope) was taken as a choice between immediate temptation versus a long-term investment that offers no immediate gratification. Spears' Indian subjects were randomly chosen to play the roles of being poor and or being rich, where the former could only have one item, be it by choice or assignment. Following this "intervention" subjects were then given performance tests using both handgrips and a modified Stroop test. The length of time that a person can squeeze a handgrip turns out (surprisingly) to be a good measure of self-control and has little to do with strength (Spears 2011, 8). In the Stroop test, subjects are shown a card with a number that is repeated a number of times. (For example the number 5 repeated three times.) The task requires overriding the automatic response, which is to note the number itself not number of times it was repeated and the tendency to conflate the two.

Spears found a statistically significant difference in performance between those in the poor group, who had to choose, as compared to the rich, who did not. That is to say, those randomly chosen to be poor performed worse than those chosen to be rich on the hand grip and Stroop tests (when the results were averaged). However, interestingly, there was no such difference between the two groups when the choices of what the poor were to get (cooking oil versus tiffin or rope) were made by the experimenter instead of the experimental subjects themselves. Moreover statistical analysis of the data supported the view that these results were produced by the effects of economic decision making, not the exertion of willpower, by demonstrating that there was still an interactive effect when the sample was restricted to the participants who selected the cooking oil. (The logic of this being that if willpower was in play, then those allowed to fulfill their temptation would not show any of the "depletion" effects. See Spears 2011, 13.)

In an even simpler second experiment, Spears offered participants the opportunity to buy soap and then tested them on the hand grip again. This time actual poor and (comparatively) rich villagers in India were the subjects, unlike the first in which participants were randomly assigned to poor and rich roles. Wealth made no difference to the hand grip results here when it was administered before the soap was offered. But when it was administered after subjects had to make the decision, the poor squeezed for 40 seconds less on average (Spears 2011, 17). Mani et al.

(2013) established similar results using cognitive function and cognitive depletion tests across four groups, comparing rich and poor subjects assigned to make expensive and inexpensive hypothetical decisions.

Lack of slack, both financial and cognitive, as well as greater hyperbolic discounting, makes the lives of the poor different from ours and makes their relationship between tradeoffs in the here and now and those in the future different as well. But in many other respects, features of all of our decision making deviate from the norms of rationality in ways that matter when it comes to the tradeoff of between economic growth now and avoiding the risk of climate change, be we rich or poor.

IV

A rational agent, operating with limited resources, will distribute those resources to maximize utility as dictated by his or her preferences. To do that, there have to be no restrictions on how resources can be distributed across the tableau of alternative "baskets" of products and services those resources can be used to purchase. In other words, resources are assumed to be fully fungible. But it turns out that is not the way we approach the marketplace. One of the most obvious places we show this is in the differential treatment of money and credit cards. Money may burn a hole in our pockets, but, by and large we are willing to pay more for something using a credit card than when we use cash (Raghubir and Srivasta 2008).

Failure of fungibility occurs more generally in a variety of contexts in which our mental accounting assigns assets as if belonging to different and nonequivalent classes (see, for example, Thaler 1990). Thus, a rational agent, treating savings as fully fungible, ought to have the same marginal propensity to consume irrespective of the source. Such an agent ought not to differentiate between current income, (known) future income and assets in hand. But in fact we tend to treat these as three different "accounts":

> Consider two professors. John earns $55,000, paid in monthly installments. Joan is paid a base salary of $45,000 paid over twelve months, and a guaranteed extra $10,000 paid during the summer months. The standard theory predicts that the two professors will make identical saving decisions. The mental accounting formulation predicts that Joan will save more for two related reasons. First, since her "regular" income is lower, she will gear her life-style to this level. Second, when the summer salary comes in a lump, it will be entered into the assets account, with its lower MPC [marginal propensity to consume].
> (Thaler 1990, 198)

Creating such artificial "accounts" is not the only way in which our thinking disrupts the basis for making a sound comparison between costs now and costs later when it comes to assessing the risks of climate change. Other factors are also in play; including, loss aversion, framing, anchoring, salience, imaginability, familiarity, and ease of recall.

Loss aversion occurs when agents value the loss of $x more than the gain $x. So, when experimental subjects are asked to set a price on an item (for example a coffee mug or a pen), they will set a higher price when they have been endowed with ownership of it as compared to setting a price to acquire ownership. That is, when asked what they would sell it for if they owned it as opposed to what they would bid on it to become an owner of it. In Kahneman et al. (1990), the seller's median price for the coffee mug was $5.25 against the bidder's median price of $2.75. While in the case of the pens the contrast was $2.06 versus $.75. Of course in any market situation, it is in the interest of sellers to set the price as high as they can, while it is in the interest of buyers to do the opposite. It is from there that market mechanisms enable the discovery of a market clearing price. But that is not what is going on here. Using tokens with experimenter assigned values if they were cashed in, but were otherwise worthless as controls to model goods without any endowment effects, Kahneman et al. were able to isolate the effect of buyer and seller market strategy from the hypothesized effects of loss aversion (see Kahneman et al. 1990, 1328).

Framing effects occur when differences in phrasing affect the way identical choices are evaluated. For example, in Tversky and Kahneman (1981) subjects were asked to choose between two treatment options for 600 people:

Under option A 200 people's lives are saved.
Under option B there is a 33% chance of saving all 600 people and a 66% possibility of saving no one.

Seventy-two percent of participants chose option A, whereas only 28% of participants chose option B. But these results were not obtained when the same choice was posed to another set of subjects, but phrased differently as:

Under option C 400 people die.
Under option D there is a 33% chance that no people will die and a 66% probability that all 600 will die.

Described this way, 78% of participants chose option D (the equivalent to option B), whereas only 22% of participants chose option C (the equivalent to option A). Notice that these two problems are identical, they are just differently described; one in terms of the lives saved, the other in terms of the lives lost. And yet that difference alone is associated with a dramatic difference in risk aversion. (You can examine your own intuitions by testing your reaction to a choice of whether to have an operation with 90% chance of success in contrast to your reactions of whether to have an operation with a 10% chance of failure. If you are like most people, you will have viewed the first option more favorably even though the risks for each is of course identical.)

Anchoring bias skews judgments as a function of an initial reference point even if it is entirely arbitrary. Thus Tversky and Kahneman (1974) asked subjects to estimate whether the percentage of African nations in the United Nations was higher or lower than a randomly assigned figure of 10% or 65%. When then asked to estimate that actual percentage, the first group set it at 25% but the second set it at 45%. Such anchoring effects are remarkably widespread and robust. Think only of how accommodation takes place to an increase in gas prices so that if there is a subsequent drop to the old price, it now seems like a "good deal." Anchoring bias not only affects estimates of size or value but also is in play in moral matters. Thus Brite and Mussweiler (2001) tested judges to see the effects of prosecutorial sentencing demands on the actual sentence. They found an 8 month difference between judges that were presented with a demand for a 34 month sentence in comparison to those presented with a 12 month sentence.

Salience is one of a class of causes of bias in assessing the likely frequency or subjective probability of an event in which the ease with which instances come to mind is treated as the basis of their likelihood. Other causes include ease of imaginability as well as simple familiarity (see Tversky and Kahneman 1974). More subtle causes are illustrated by differences in the ease in which we can complete some tasks over others. Tversky and Kahneman (1973) asked subjects whether a randomly drawn word from an English text of three letters or more was more likely to start with the letter "r" than it being the third letter. They argued that people decide this by recalling words that conform to each of the conditions, but because it is easier to recall words that start with the letter "r" than those with the letter in the third position, their assessments are systematically unreliable.

Modeling decision making with reliance on the assumption that our reasoning conforms to the axioms of rationality represents a compromise between verisimilitude and mathematical tractability. But in cases where the latter is not needed the verisimilitude we give up by means of such an assumption extracts an unnecessary price in such models and obscures understanding of how bad we are at making certain decisions, as the foregoing have illustrated. Weighing the choice between foregoing economic growth now and avoiding climate change in the future is especially vulnerable to these features of our decision making because it involves a choice between the familiar in the here and now and the unfamiliar in the indeterminate future. That makes the choice especially vulnerable to the biases we have been considering:

> Mental accounting leads us to differentiate between current and future income.
> Framing induces us to weigh potential gains (in current income) differently from potential losses (due to climate change).
> Anchoring provokes parochialism in our assessments by tying it to our current level of wealth.
> Salience, imaginability, familiarity, and ease of recall all favor over-valuing the chance of short term gains over the risk of long term losses.

Finally, although the choice does not involve giving up something we have now for something in the future, loss aversion still engages when we imagine giving up something we could have now (see Mellers 2000).

V

Even if these biases are universal among us, the circumstances of decision making are not, and make for a difference again between the poor and the rest of us. Lack of slack, as we have seen is a two-edged sword. It forces greater attention to trade-offs between choices in the here and now but in doing so limits the (cognitive) resources available for attending to tradeoffs between the here and now and the future. Moreover there are other psychological elements of the circumstances of the poor in play as well, like low literacy and the associated stigma of it. Adkins and Ozanne (2005, 104) report that "When consumers accepted the stigma of low literacy, market interactions were perceived to be risky because one's identity was vulnerable to potential assaults." Moreover, as Barr (2012, 4825–4832) reports the poor are more likely to find themselves in situations that are distracting and tense which in and of themselves consume cognitive resources and degrade decision making.

But if this suggests a potentially unsystematic approach to decision making that such circumstances can precipitate, one consistent feature of decision making that we have seen is that poverty is associated with an increased aversion to risk taking. Now you might think that when it comes to avoiding climate risk, generalized risk aversion ought to work in its favor. But I want to argue that in this calculus, even if the poor are more risk averse than the rest of us, their circumstances undermine this consequence.

For most of us there is a meaningful tradeoff between reducing the well-being of our immediate offspring to reduce the risk of harming future generations, all the more so if they are genetically related. But notice also the trade is not all or none for us. Our wealth means that the reduction we might visit on our immediate offspring will only slightly reduce their well-being. Maybe they will have smaller cars and houses than we have had. Or maybe they will have to live in denser urban settings close to work. Or maybe they will not be able to heat and cool their homes with the sense of abandon that we have had the luxury to do in our lifetimes. But these are trivial sacrifices. None of them endanger the survival of our offspring. That would make no sense. If your immediate offspring don't survive, nor do future generations.

But consider this tradeoff through the eyes of the world's poor. If you are poor, survival of your offspring is really in question in a way that it is not for you and me. For them, there is no tradeoff in the sense that there is for us, since (as we noted earlier) the road to the survival of future generations goes through the road to the survival of their immediate offspring. So even if the poor have a higher degree of risk aversion than the rest of us, that fact is unlikely to engage when it comes to assessing the risk of causing climate change as a consequence of economic growth.

VI

Hemingway is responsible for a famous misquotation of Fitzgerald's about the rich. "The rich are different than you and me – they have more money." Hopefully the same idea is true in reverse of the poor – they are only different in that they have less money. If that is right, the key question is how much economic growth is needed to make the choices of the poor less stark by making them less poor. And with it what the climate change consequences are. But that said, rich or poor we are all subject to the vicissitudes and idiosyncrasies of the way in which we assess and value choices.

Moreover, even if we were totally rational agents, there is another way our decisions can be distorted in the face of choices we face individually and socially in the face of what might be early signals of climate change. Whether you think the incidence and intensity of hurricanes, the number and severity of tornadoes or the frequency and range of forest fires are a product of climate change does not matter for the purposes of this discussion. For, whether related or not, they offer test cases for the way we approach the distribution of risk collectively and the way in which that affects the rationality of our decision making individually.

References

Adkins, Natalie and Julie Ozanne, 2005. "The Low Literate Consumer," *Journal of Consumer Research*, 32: 93–105.

Barr, Michael 2012. *No Slack: The Financial Lives of Low-Income Americans*, Washington DC: Brookings Institution Press. Kindle Edition.

Bertrand, Marianne, Sendhil Mullainathan and Eldar Shafir 2006. "Behavioral Economics and Marketing in Aid of Decision-Making among the Poor," *The Journal of Public Policy and Marketing*, 25 (1): 8–23.

Brite, Englich and Thomas Mussweiler 2001. "Sentencing under Uncertainty: Anchoring Effects in the Courtroom," *Journal of Applied Social Psychology*, 31: 1535–1551.

Family Feud & Friends 2012, http://familyfeudfriends.arjdesigns.com/ accessed October 10th, 2012.

Kahneman, Daniel, Jack Knetsch and Richard Thaler 1990. "Experimental Tests of the Endowment Effect and the Coarse Theorem," *Journal of Political Economy*, 98 (6): 1325–1348.

Lewis, Michael 2012. "Obama's Way," *Vanity Fair*, October, www.vanityfair.com/politics/2012/10/michael-lewis-profile-barack-obama, accessed October 14th, 2012.

Mani, Anadni, Sendhil Mullainathan, Eldar Shafir and Jiaying Zhao 2013. "Poverty Impedes Cognitive Function," *Science*, 341: 976–980.

Meier, Stephan and Charles D. Sprenger 2011. "Time Discounting Predicts Creditworthiness," *Psychological Science*, 23 (1): 56–58.

Mellers, Barbara 2000. "Choice and the Relative Pleasure of Consequences," *Psychological Bulletin*, 126: 910–924.

Mullainathan, Sendhil and Eldar Shafir 2009. "Savings Policy and Decision Making in Low-Income Households," in Michael Barr and Rebecca Blank (eds.), *Insufficient Funds: Savings, Assets, Credit and Banking Among Low-Income Households*, New York: Russell Sage Foundation Press, 121–145.

Pryer, Jane 2003. *Poverty and Vulnerability in the Dhaka Slums*, Aldershot: Ashgate.

Raghubir, Priya and Joydeep Srivastava 2008, "Monopoly Money: The Effect of Payment Coupling and Form on Spending Behavior," *Journal of Experimental Psychology: Applied*, 14 (3): 213–225.

Sachs, Jeffrey 2005. *The End of Poverty*, London: Penguin.

Shah, Anuj, Eldar Shafir and Sendhil Mullainathan 2012. "Some Consequences of Having Too Little," *Science*, 338: 682–885.

Spears, Dean 2011. "Economic Decision-Making in Poverty Depletes Behavioral Control," *The B.E. Journal of Economic Analysis & Policy*, 11 (1), Article 72.

Thaler, Richard 1990. "Saving, fungibility and mental accounts," *Journal of Economic Perspectives*, 4: 193–205.

Tversky, Amos and Daniel Kahneman 1973. "Availability: A Heuristic for Judging Frequency and Probability," *Cognitive Psychology*, 5 (2): 207–232.

Tversky, Amos and Daniel Kahneman 1974. "Judgment under Uncertainty: Heuristics and Biases," *Science*, 185: 1124–1131.

Tversky, Amos and Daniel Kahneman 1981. "The Framing of Decisions and the Psychology of Choice," *Science*, 211 (4481): 453–458.

Wood, Geoff 2003. "Staying Secure, Staying Poor: The Faustian Bargain," *World Development*, 31 (3): 455–471.

6

SOCIAL POLICY AND RATIONAL ACTION

I

A few weeks after Hurricane Sandy, I drove down the coast of New Jersey to see the damage firsthand. As I drove from the inland highway through beach towns toward to coast, bags of fallen leaves waiting for collection on the sidewalks gave way to bags of sodden insulation material from basements that had been flooded. In most towns basement flooding was the most common damage more than two or three blocks from the ocean. Closer than that though and I could see houses wearing tarp caps to cover bald spots on their roofs and others with missing siding and Tyvek showing like underwear. But none of that prepared me for the scene by the ocean itself. I followed a convoy of home repair vans through the town of Brick until we reached the water. Forlorn homeowners were still picking over the remains of their possessions arrayed outside their damaged houses, many of which were marked for demolition. Up and down the coast, bridges were down, boardwalks destroyed, and the sand on beaches sat in gigantic piles with bulldozers already at work redistributing it. That this was no aftermath of a tsunami that reached far inland made the dollar estimates of the damage of almost $37 billion all the more astounding (Dopp 2012).

The last time I had been in Ocean Grove was when my kids were still kids. Twelve years ago I remember warning them not to climb on the dunes whose fragile grass was posted with warning signs. If it comes as a shock to see the natural landscape so easily "de-constructed" by the storm, it comes as more of a shock to realize that what seemed so natural before was itself constructed. Those fragile dunes were the result of a man-made process over 25 years in which over $700 million were spent moving sand on about 55 % of New Jersey's shores (Farrell 2013).

That should give one pause about the likely size of the bills in the future if my colleagues in climate science at Rutgers are right about the likely increase in the frequency and intensity of storm surges (see Miller et al., forthcoming). The associated increase in the rate and severity of beach erosion means new sand will be needed, on average, every 4 years at a cost of over $850,000 per mile per year, based on the estimated cost of $400 million to repair New Jersey's 120 miles of developed shoreline in the aftermath of hurricane Sandy (as estimated by The Program for the Study of Developed Shorelines at Western Carolina University).

An island off the coast of Alabama may become the norm. The housing and infrastructure on the western end of Dauphin Island has been decimated by a dozen hurricanes and storms since 1979, only to be rebuilt with federal support (Gillis and Barringer 2012).

A combination of federally subsidized flood insurance for homeowners and FEMA emergency grants, after the fact, shield communities like Dauphin Island from the true costs of rebuilding in vulnerable areas. These subsidies create two grades of disincentives for these communities to look nature in the eye. At the first level, the cost of flood insurance provided through the National Flood Insurance Program (NFIP) is subsidized by all taxpayers with about $200 million annually (GAO 2003) and hence those who do not need it subsidize those that do. At the second level, the distribution of disaster aid to both local governments and citizens requires no insurance and thus constitutes a prima facie case of moral hazard. Let us take these up one at a time.

II

Suppose we begin with the simple idea that it is useful for a population at equal risk of suffering equal loss to share the total cost of the loss equally among themselves, unless that risk is a certainty for all of them. So, in a population of n people, m will suffer a loss of $x, so the premium is $m.x/n. The same idea makes sense for a population in which the risk may not be equally distributed among its members but the distribution of risk is unknown. Now suppose that the distribution of risk is known and is known not to be equal. All other things being equal should that not be reflected in differential premiums? Our practice in this regard is quite complicated and nuanced.

For example, we know that young drivers are at a higher risk of causing accidents than older drivers and we reflect that in a differential premium. But youth is not the only way of differentiating the risk pool. Yet, in California, under the provisions of Proposition 103, the only additional factors insurance companies are meant to take into account in rate setting are the insured's driving safety record and the number of miles driven annually. Attempts by insurance companies to challenge this provision in favor of the use of additional factors with greater alleged predictive power of loss were not upheld by the courts. Instead Proposition 103 only allows additional factors to be used in rate setting "as the commissioner may adopt by regulation that have a substantial relationship to the risk of loss." To date, the most controversial of these has been the use of zip codes as promulgated in 1997 by the California Insurance Commissioner and upheld on appeal.

Suppose zip codes are a good predictor of loss. But suppose there are other good predictors as well. Suppose some neighborhoods not only have a higher auto theft rate than others but are also correlated with a higher probability of being found at fault in an accident. Of course once we set off down this road, depending on how much we are willing to fine grain the data, many other factors may end up as statistically robust predictors of loss. Suppose it turns out that right-handed drivers have a higher accident rate than left-handed drivers in the United States. Nobody knows why this should be the case although theories abound. Some think it is because left-handed drivers more naturally turn their head to the left and are thus more likely to check their blind spot. Others think it is because left-handed drivers favor their left legs for braking and are thus quicker to brake than right-handed drivers. But whatever the reason, suppose that the statistical finding turns out to be robust.

If you think of insurance as a voluntary compact that people enter into of their own choice with the ability to bargain for pricing, it does not seem unreasonable for left handers to demand lower premiums for right handers. Nor does it seem unfair. Under what circumstances, if any, might it be unfair? Consider this: there was a time when left-handed people were forced to learn to use their right hand to write with. They became involuntarily right handed. Suppose with that they also became less fluid in their movements. Suppose the class of right-handed drivers is composed of the naturally right handed and those forced to become right handed. And suppose the data showed that it is this latter class, not the former class, that accounts for the statistical difference in the accident rates between left- and right-handed drivers. Now suppose you are a natural right-handed driver, but stuck with a higher premium than me, a natural left-handed driver. You might well complain that you are being forced to carry a risk premium that you do not deserve to carry. But notice that complaint is only going to find a footing if we have the data available that allows us to separate out the formally left-handed right handers from the rest of the right handers. And if we have the data available to ground such a complaint, of course, there is nothing to prevent us differentiating the rates to reflect that data. Indeed the more we care to fine grain the data, the more we have a chance to fine grain the rate structure of premiums.

Except for the time and money, is there any reason to object to doing that in principle? As Dicke (2004) argues, the more we can do that the more efficient the pricing of insurance becomes overall. The reason is that fine graining the rate structure goes hand in hand with greater accuracy of risk assessment for the totality of the insurance pool that reduces the risk premium associated with ignorance. But more importantly, *failure* to offer differential premiums can lead to the collapse of insurance schemes as the less risky insured leave the pool, concentrating the risk and raising the premiums of those remaining.

Still, even if differential pricing is necessary to ensure that a market insurance scheme succeeds and it promotes overall price efficiency, that is not to say that some people may not face impossibly high premiums if they fall into high risk categories. As Dicke (2004, 67–68) points out, the most dramatic of these are those seeking life insurance with a diagnosis of Huntington's disease. Huntington's is a rare instance in which the presence of a genetic mutation is an infallible indicator of the inevitable occurrence of the disease and life expectancy is generally 20 years after the onset of initial symptoms (Walker 2007).

The idea that someone with Huntington's disease deserves coverage, and coverage at an affordable price, is the most extreme instantiation that, contrary to the claim above, relative pricing as a function of relative risk may not in fact be morally fair. Bracketing the question of how an insurance market that does not set prices relative to risk can work for a moment, the intuition here taps into the same vein of luck egalitarianism we examined earlier in a different context. That is the idea that "[a]mong equally deserving people, it is bad, because [it is] unfair, for some to be worse off than others through no fault or choice of their own" (Temkin 2011, 51). Here the issue is what risks (if any) should fall under the umbrella of this principle. If there is a case to be made for anything it is surely genetic inheritance. Indeed Norman Daniels (2004, 125) thinks it applies much more broadly, in opposing the idea that "individuals are entitled to benefit from their individual differences," especially when it is a matter of lowered risk for disease and disability. What this comes down to saying is that those that lack such benefit as a result of their individual differences are entitled to help to make up for those differences wherever possible. Of course that is not to say we should oppose a benefit derived from any and all differences. But as Daniels points out (2004, 128), it is only the strictest of libertarians that would include differences in the rates of disease and disability on this list.

III

If we recognize the notion that disease and disability should not be the basis for differential premiums, to what other arenas of insurance, if any, should this idea be extended? And in particular, should it be extended to the areas of our current concern, coastal flood insurance? Daniels (2004, 121) defends what he terms a "multifunction thesis" about insurance. By his lights, medical insurance is different from fire insurance. Fire insurance is simply a tool for risk management. On the other hand, medical insurance not only plays that role, but is, Daniels thinks, also a vehicle for the discharge of a social obligation. We need to examine the contrast between these two functions.

For Daniels the crucial social function in play in medical insurance is that we must protect "equality of opportunity by assuring everyone access to necessary medical services." From that he concludes that "justice requires that we not permit risk exclusions or risk rating, or actuarial fairness, to constitute barriers to insurance coverage and thus to access to care" (Daniels 2004, 133–134). But of course rate differentiation and actuarial fairness per se don't constitute a barrier to access. After all, not every degree of rate differentiation is going to be large enough to prevent access to medical services.

We might take Daniels' starting point of protecting equal opportunity to medical care and interpret "equal" literally, arguing that rate differentiation in and of itself undermines the equality of opportunity, even if it is not a barrier to access to care. But is equal opportunity really what is at stake here? When we say race, or gender, or genetic inheritance shall not be a basis for differential rates, it is out of principle that we are carving out arenas to be off limits as a basis for prediction. In doing so, we are asserting that these are not the sort of categories that we want to treat in a means-ends way, not because of the effects on access such a differentiation might. And so too when it comes to, at least, some medical conditions. There the principle may seem pretty straightforward: we do not want to take account of things outside people's control.

But as Hellman (1997) points out, control per se may not be a reasonable condition on its own. There are some conditions we might be able to control, but doing so would be too burdensome, as in the hypothetical she offers of someone who must not move to maintain his health. The notion of burdensomeness is of course hard to pin down, but however we choose to do so, it is always going to be relative to some (normative) standard of "normal" functioning. Moreover, with or without an appeal to the notion of burdensomeness, the notion of control itself is fraught with difficulty if we pay the least attention to the havoc that weakness of the will can create, at least for most of us.

Is there anything more we can say about the basis for deviating from risk differentiated premiums? A popular way to try to do so is in terms of the concept of solidarity. Launis (2003) explicates this idea in terms of a notion of reciprocity in contrast to charity or mercy. As he puts it:

> In the state of reciprocity (mutual responsibility), the community is obligated to stand by needy individuals . . . just as the individual is obliged to stand with other community members.
>
> (Launis 2003, 92)

That is to say, in solidarity, people are treated equally and have a mutual right to expect help from each other.

But this seems problematic to me in two different respects. First, as Lehtonen and Liukko (2011) argue, the very idea of insurance as risk management is itself rooted in notions of solidarity that underwrote the rise of insurance programs in the 19th century that replaced the idea of individual responsibility. Relying on the work of François Ewald, they argue that the widespread use of insurance arose with the idea that "the responsibility for accidents should be shared by all members of society, because their cause was essentially social" (Lehtonen and Liukko 2011, 35). On this view then, the idea of solidarity is already baked into insurance and we need to distinguish between two kinds of solidarity, what they call "chance solidarity" as opposed to "subsidizing solidarity."

Are we involved in anything more than a semantic squabble here? I think we are because the idea of solidarity as understood so far does not really do justice to just what is involved when it comes to the subsidizing role of insurance; namely that far from being mutual or reciprocal it is, in fact, asymmetric. When someone is known to have the genetics associated with the inevitable onset of Huntington's disease and is given life insurance nonetheless, there is no reciprocity. Instead, solidarity expresses itself in our commitment to take care of that person. In this sense, contra Launis, solidarity here *is* akin to a form of charity or mercy. Thus chance solidarity and subsidizing solidarity are quite different from each other. But if the case of being at risk for Huntington's disease is an instance of this, it is an instance of a widespread phenomenon of the use of insurance as an instrument of social policy as well as risk management. Nor is it the most extreme instance of it. For that consider government regulations that prohibit the exclusion of people with preexisting conditions (as opposed to those merely at risk for those conditions) from gaining insurance, which essentially functions as a redistributive instrument unrelated to any notion of risk management.

Now you might object and try to argue that risk management does enter in at a different level in the exercise of these policies. For all I know, I might be diagnosed in the future with some newly discovered genetic compliment that spells my early demise. Or I might want or need to change insurance notwithstanding a preexisting condition. So, using insurance as a social policy might itself offer an offset against the risk of these states of affairs that might affect me adversely. But this seems a stretch since we use insurance as a social policy to provide coverage to people at risk for outcomes that some providing the subsidy could not face. Thus, for example, the state of New Jersey mandates that insurance companies must provide post-natal inpatient care for mothers for at least 2 days after a vaginal delivery (and at least 4 days after a Cesarean delivery). The cost of this benefit is borne by all. But this is not risk management for a pool of the insured since only some of the members of that pool (women) are at "risk" of giving birth. Instead what we have here again is a case of subsidizing solidarity without reciprocity, at least when it comes to men who presumably face no risk of pregnancy.

Now one might object as follows: of course there can be no reciprocity in the strict sense of a pool of those all at risk for pregnancy if it includes men. But reciprocity needs to be understood more broadly. By extending coverage to those in need at subsidized rates, I guarantee that I too will benefit from subsidized rates were I to fall into a high risk class. But there are two problems with this way of looking at things. The first is that if we understand a subsidy in that way, there ought to be equivalency with how we price the known risk someone faces and the unknown risk that I may face. Rate equity means that we treat these as of equal value. But we clearly don't. For if we did, you ought to be indifferent between these two states of affairs – one in which you face a known risk (like Huntington's) and one where you don't (but for all you know might come to do so). Moreover there is another pretty obvious reason why we don't value them the same. For just because a person has Huntington's doesn't preclude his being at risk for other unknown bad outcomes no different than the ones you are worried about. In this sense the person with Huntington's will always be at higher total risk than a person without it, and hence it would be irrational to choose the former over the latter if given a choice.

That said, there is a much more powerful reason to think there should be no hidden reciprocity in play that we can find. For what our practice reflects is a social commitment to support those in a different risk category from the norm in the same way as we treat others, as a matter of principle. The principle is no different than what drew us to drawing straws in a lifeboat to decide who would eat and who would be eaten. In that situation, by drawing straws we choose to treat people as deserving an equal chance irrespective of the (possible) differential costs and benefits of doing so.

Notice however that whenever we allow social policy to override rate differentiation, achieving that goal will usually require government action to avoid the implosion of an insurance market that results when those paying more than "their fair share," qua risk, flee that market. Under these circumstances the government has three options. One is to mandate everyone stay in the pool by law. Another is to regulate all insurance markets to ensure that uniformity when social policy is to be allowed to trump rate differentials. The third is to leave the rate differentials in place, but simply reimburse those paying more out of general funds.

However we choose to safeguard the integrity of insurance markets when social policy trumps other considerations, it is worth noting that in the lifeboat case, under some circumstances, I claimed most of us could be pushed to letting consequentialist considerations override our straw drawing sentiments. As in the lifeboat case, here too we can ask under what circumstances we are inclined to blinker ourselves to risk differentials and where we are not, and why.

You might think that the straightforward answer hinges on the balance of control and burdensomeness that we broached earlier. Those who ride motorcycles should not be subsidized by those who drive cars. And those who don't wear helmets should not be subsidized by those that do. On the other hand, those with predispositions to diseases over which they have no control should be subsidized. But even these seemingly clear cases are not that clear. What if people are too poor to drive cars but can afford (low-powered) motorbikes? And what of parents who are counseled that any children they have will have a 100% chance of having a disease that is costly to treat?

These kinds of examples suggest there will be no easy way to draw a line between the kinds of circumstances that we want to exclude from playing a role in rate setting and those that we want to allow to play such a role. But that said, there is nothing to prevent us from sorting clear cases that merit different treatment. With that in mind, we are finally ready to ask the question of whether or not insurance for those who live in areas prone to the effects of climate change, like increased coastal flooding and hurricane damage, or increased wild fires, constitute a class of clear cases or not.

IV

As the case of Dauphin Island attests, our current policy is to insulate coastal residents from the full costs of insurance coverage. Moreover, in some cases (like coastal Florida) where the insurance market won't provide coverage because of the level risk, the state itself has stepped in to provide coverage. So we blunt the effect of market forces that would drive up the cost of coastal living or even make it impossible for those needing insurance as a condition of obtaining a mortgage. Now it might seem obvious at first blush that this is an irrational policy. It would be one thing to allow people to freely choose to live where they want and bear the full cost of that choice, even as that cost goes up as nature extracts a higher and higher price. But why should the rest of us subsidize that choice any more than we should subsidize the choice of a bungee jumper or a mountain climber? Now in one sense we do in fact subsidize these last activities. Not by insurance perhaps, but by providing rescue and medical services when things go badly for them, without requiring them to cover the full cost of such services. Even if we decide to offer the same rescue services to those who live in coastal regions prone to flooding or hurricanes, and perhaps we should throw that into question as well, why should we do so by way of insurance subsidies?

It is important to emphasize the "should" here. Local elected representatives of coastal communities affected by climate change will certainly try their best to obtain such subsidies, and the realities of political quid pro quos may make it quite likely that they will succeed. But should they succeed? All other things being equal you might say the default argument is that they should not. Indeed the default position may be even stronger than that to the extent that people's free choices inevitably place costs on the rest of society. So even if people bear the full cost of building homes in coastal areas at heightened risk of damage due to climate change, the minimum services a state is expected to provide to all of its residents like police, fire, and postal services create a burden we all bear equally. At least it does unless we institute differential fees for these services, which is not something we do in the course of normal business. A first-class letter costs the same to send notwithstanding the vastly different cost of sending it across town in New York as compared to sending it from New York to zip code 99685 in the Aleutian Islands.

Living in the Aleutian Islands is certainly not easy. The 4,376 residents of Unalaska live only 13 feet above sea level. So in the long run they will be a coastal community at risk for more than flooding and likely require complete evacuation. But between now and then, as one of the most isolated communities in the United States, located in some of the harshest terrain, the whole enterprise of maintaining a community there is very expensive. For example the nearest hospital is 462 miles away and the ambulance service is by air. And with the nearest town with a population center of more than 50,000 nearly 800 miles away, the provision of many government services entail a substantial premium for travel that is subsidized by other tax payers.

Unalaska is an extreme example of how much the costs of services can deviate from the mean. But the fact of the matter is that almost all communities will deviate from the mean in one respect or another, reflecting differences in the cost of the delivery of services. For some services, rural settings cost more because of low density and distances (like mail delivery). For other services, the higher population density in urban settings increases the demand on services (like public health oversight of infectious diseases). But we don't treat these differential costs as appropriate to reflect in differential taxes or user fees. Still you might think that is a matter of practicality rather than principle. It would be time consuming and costly to calculate and collect these differential costs.

The same issue arises much more starkly when it comes to who pays and who benefits from government support after natural disasters. Some people live in Tornado Alley and some don't. Some people live near or even on earthquake faults and some don't. Some people live in areas prone to wild fires and some don't. Here the calculation of costs is much easier to make because all we have to do is add up the disbursement provided under federal relief. But of course the idea of a federal *relief* program is that we think it right as a matter of principle for us to all share the costs of the disasters that befall some. If we think of this as an insurance program from which we all stand to benefit, it is nonetheless an insurance program in which we all pay the same rate even though we are not all subject to the same risk of disaster. Is there another way to think of such a program? It can't be that we face unknown risks wherever we live because as we saw earlier, so too do those who live in areas of known risk. Instead perhaps the reason is this: we want some people to live in risk prone areas and we are willing to reward them for doing so. Better them than us we say, just as we do when we offer hazardous duty pay for some professions.

Here is a case where this model is applied for better or worse. When the Labor Party of Israel gained the West Bank after the 1967 War, they wanted to encourage settlement of the new land to both claim it and (purportedly) to discourage border incursions from Jordan. Housing subsidies were granted to the settlers as a reward for serving what was claimed to be the national interest. But it is hard to see how such an argument applies in the United States. You might say, it is in our interests for our coastal regions to be populated for commercial reasons. But that would only ground subsidies for coastal commercial operations and (perhaps) their workers. When it comes to places like New Orleans that is only a very small proportion of the population.

Instead it seems much more plausible to say, for better or worse, we have settled communities in places that are more dangerous than others and it is not fair to uproot those communities. Fair or efficient? Suppose the risk of where they live arose after the communities were established. So there can be no question that they knowingly chose to populate an area at high risk. Even then, is the matter one of fairness? Those committed to the indigenous culture of New Orleans would certainly say it is. To move is to lose their unique way of life. We, who do not live in New Orleans, may say we value their life so much we will pay to preserve it. But if we don't feel that way, is it unfair for them, and to them, for them to lose it? If there is a claim here, it does not strike me as an overriding claim or a claim that can't be trumped by other considerations. The energy needs of a society may require damming a river that entails submerging some communities. They will lose their way of life. We surely have an obligation to compensate them and to offer them a place to reconstitute their way of life to the extent that is possible. But as we saw in other contexts earlier, even where we can't do this, it does not rule out welfare considerations outweighing these concerns.

In the aftermath of Hurricane Sandy, numerous people who lived by the sea in New York in Staten Island and Brooklyn declared their setting to be something they would never want to give up despite the increasing hardship of living there. But the fact they were lucky enough to live there, often over many generations, creates no basis for which they may be said to possess a case for a claim against the rest of us for detrimental reliance. Or does it? They may say, "Look in the past you (the state) helped fix flood damages, just as you fixed damages in other communities caused by fire, earthquakes, landslides and so on, and we have planned our lives accordingly." That is a fair argument for help to move to safer ground perhaps. But it does not sanction an obligation for the state to subsidize the status quo.

"Hold on!" these coastal residents may object, "If we are to bear the full cost of the risk of living by the sea, then at least make others bear the risk of where they live: those in Tornado Alley, those on the San Andreas Fault, those in high risk fire zones in the Southwest, or those near Hawaii's Kilauea volcano."

This complaint makes stark the inelegant fit between considerations of equity and considerations of efficiency. On the one hand we have a commitment to subsidizing differential risk that pulls in one direction. On the other hand, at some point that subsidy can reach a large enough size that another interest trumps it: it becomes a matter of social interest to discourage people from certain risk settings. The problem is there is no smooth way of combining these interests without subsidizing the less impacted more than the more impacted in a schedule that moves seamlessly from subsidies to penalties.

V

If insurance subsidies pose policy dilemmas, they fade in comparison to our practice of providing some forms of disaster relief *after* the fact to those who have no insurance in the first place. This offers the purest instance of moral hazard that undermines an incentive for those exposed to higher risk than others to move. If you lack flood insurance and your house is flooded you must suffer the consequences of your lack of foresight. So too if your lack fire insurance and your house burns down, or you lack earthquake insurance and your house collapses – unless these disasters befall you in what the federal government declares a federal disaster area. Then your lack of foresight will be partially offset with low-interest loans to rebuild, as well as government aid for temporary housing, transportation, medical costs and even funeral costs.

These are programs that treat the same number of people suffering the same level of disaster differently depending on whether they are in one place or not. One hundred single home fires in many communities do not qualify, but one fire of 100 houses in one community can qualify. The underlying rationale for this is that a state or locality might be able to handle the first but be overwhelmed by the second. Hence federal assistance is reserved only for the second. But of course state policy does not act in harmony with federal policy in taking care of individual losses even if states could afford to do so. For that would create an obvious further moral hazard. This creates an unintended double standard. You suffer a loss in isolation without insurance you are on your own. But if you suffer the same loss in the company of others, your lack of foresight does not get (fully) counted against you.

That may seem unfair, but what we have here again is the jagged transition when considerations of welfare trump other considerations. When the numbers get large enough, they count not just in trumping considerations of fairness but in overriding concerns about moral hazard as well. But is moral hazard the same when the numbers are large as when they are small? When it comes to fire, you might reasonably think that the cost of this moral hazard is limited in the sense that the usual fire risks we face are usually visited upon us one at a time and only occasionally in the company of others. One person's house fire is not normally coincident with all of his or her neighbors. So if you are going to balance the cost of the moral hazard against the benefits of general welfare in a disaster that affects many, moral hazard may be more restricted than you think. Providing after-that-fact disaster relief for a neighborhood burnt down by a fire is not going to discourage people from buying insurance to cover their house burning down alone. And not just fire. For example, FDIC insurance does not normally cover money market accounts. But in 2008, the U.S. Treasury Department provided 1 year of insurance for these funds to cover losses of up to $50 billion because of the widespread threat to the financial system. Now you might reasonably expect that a rational agent should think that such insurance will be reinstituted under the same circumstances, but that should not give rise to an expectation of help in the much more likely occurrence of the looming failure of an isolated money market fund.

But notice how this division of risks and moral hazard does not obtain for some risks. Unlike fires, hurricanes never affect just one or two individuals. They are by their nature large scale events. So too are floods. But these are just the kinds of risks that are characteristics of the effects of climate change. When they have effects, their effects are usually widespread and implicate many people. So the calculus of moral hazard and welfare here is different. These programs are always at odds with moral hazard. They have no arena that is insulated from its reach, unlike disasters that can range in their scale of visitations from just one household to a vast number.

Helping communities in the aftermath of hurricanes, while encouraging them to face the risk of climate change as it is priced into the insurance market, thus pull in different directions. The first interest threatens to undermine the second. And while helping communities in the aftermath of a mass fire, while encouraging them to buy insurance does so as well, in the case of the latter, the prevalence of individual fires blunts that effect. But that is not to say resolving to remove incentives that reinforce the status quo would need be implemented without reasonable notice nor need they be implemented in an all or none method. Thus a policy to stop disaster aid or underwriting insurance in areas designated as too risky to populate could be announced with a two decade notice. Moreover during that time, continued aid could be conditioned on rebuilding only in new areas away from danger that is already part of a plan in New York in the aftermath of Hurricane Sandy. (See Kaplan 2013.)

Still what makes such policies hard to implement is deeply embedded in our electoral system. In a system of proportional representation, geographically specific interests become diluted by broader considerations. However in a system like the United States in which representatives are elected by districts, as alluded to earlier, narrow political interests work against policies that may be unwelcome by residents of some districts, let alone result in a diminished population (and with it voters) for that geographical area.

Overcoming these kinds of hurdles may be a sine qua non for voters to embrace the necessity for action to not just offset the effects of climate change but also to limit it. But doing that requires more than a national consensus. International cooperation is required to make a real difference. Yet many take it that the Tragedy of the Commons casts doubt on how likely this is to happen.

References

Daniels, Norman 2004. "The Functions of Insurance," in Mark Rothstein (ed.), *Genetics and Life Insurance*, Cambridge: M.I.T. Press, 119–145.

Dicke, Arnold 2004. "The Economics of Risk Selection," in Mark Rothstein (ed.), *Genetics and Life Insurance*, Cambridge: M.I.T. Press, 49–72.

Dopp, Terrence 2012. "Christie Says New Jersey Sandy Damage Now $36.8 Billion," *Bloomberg News*, www.bloomberg.com/news/2012-11-28/christie-says-new-jersey-s-sandy-damage-rises-to-36-8-billion.html, accessed January 20th, 2013.

Farrell, Stewart 2013. "The impact and history of NJ State sponsored beach restoration since 1982," *Northeast Beaches Conference*, Galloway, NJ, September 9–11, 2013.

GAO 2003. "Challenges Facing the National Flood Insurance Program," *GAO-03-606T*, www.gao.gov/products/GAO-03-606T, accessed March 18th, 2013.

Gillis, Justin and Felicity Barringer 2012. "As Coasts Rebuild and U.S. Pays, the Critics Ask Why," *New York Times*, November 18th, www.nytimes.com/2012/11/19/science/earth/as-coasts-rebuild-and-us-pays-again-critics-stop-to-ask-why.html?emc=eta1, accessed February 2nd, 2013.

Hellman, Deborah 1997. "Is Actuarially Fair Insurance Pricing Actually Fair?: A Case Study in Insuring Battered Women," *Harvard Civil Rights-Civil Liberties Law Review*, 32: 355–411.

Kaplan, Thomas 2013. "Cuomo Seeking Home Buyouts in Flood Zones," *New York Times*, February 13th, www.nytimes.com/2013/02/04/nyregion/cuomo-seeking-home-buyouts-in-flood-zones.html?pagewanted=all&_r=0, accessed February 21st 2013.

Launis, Veikko 2003. "Solidarity, Genetic Discrimination, and Insurance: A Defense of Weak Genetic Exceptionalism," *Social Theory and Practice*, 29: 87–111.

Lehtonen, Turo-Kimmo and Jyri Liukko 2011. "The Forms and Limits of Insurance Solidarity," *Journal of Business Ethics*, 103: 33–44.

Miller, Kenneth, Robert Kopp, Benjamin Horton, James Browning and Andrew Kemp Forthcoming. "A Geological Perspective on Sea-Level Rise and Impacts along the U.S. Mid-Atlantic Coast."

Temkin, Larry 2011. "Justice, Equality, Fairness, Desert, Rights, Free Will, Responsibility, and Luck," in Carl Knight and Zofia Stemplowska (eds.), *Responsibility and Distributive Justice*, Oxford: Oxford University Press.

Walker, Francis 2007. "Huntington's Disease," *The Lancet*, 369 (9557): 218–228.

7

THE TRAGEDY OF THE COMMONS REVISITED

I

The challenge of climate change is widely held to be an obvious instance of Garrett Hardin's Tragedy of the Commons par excellence:

> Picture a pasture open to all. It is to be expected that each herdsman will try to keep as many cattle as possible on the commons. Such an arrangement may work reasonably satisfactorily for centuries because tribal wars, poaching, and disease keep the numbers of both man and beast well below the carrying capacity of the land. Finally, however, comes the day of reckoning, that is, the day when the long-desired goal of social stability becomes a reality. At this point, the inherent logic of the commons remorselessly generates tragedy.
>
> (Hardin 1968, 1244)

It is this "inherent logic" of it that "remorselessly generates tragedy" (Hardin 1968, 1244). Hence the basis for widespread pessimism about our current circumstances. There are certainly many reasons to be pessimistic about our circumstances, but is the Tragedy of the Commons really one of them? It is not the logic that I think deserves reexamination, but rather its application to the case of climate change. At first blush this may seem quite ludicrous – if that logic does not apply to climate change, then what does it apply to? But in what follows I want to show how parties can step outside that logic by drawing on interests that it ignores. I first look at the application of Hardin's argument to groups of individuals and then to state actors. The widespread but erroneous belief that climate change is an instance of the Tragedy of the Commons threatens to limit our horizon of possibility, but that is not to say that cooperation is easy to win.

Garrett Hardin's famous essay is now so closely associated with the problem of regulating greenhouse gasses that it easy to forget that the essay itself hardly touches on the subject despite his iconic invocation of the pasture. The Tragedy of the Commons is primarily an essay about population growth and its discontents. It is a general meditation on the destructive consequences of increased population size and the challenge of regulating those consequences. The intended contrast is to Adam Smith's "Invisible Hand." Hardin's thesis is that individuals pursuing their self-interest, far from promoting the public interest, will in fact produce an outcome that is suboptimal for everyone, including themselves, absent unlimited resources. If at the time the essay was written (in 1968), it was population growth that seemed to represent the most likely basis for such increasing demand, today, we see it as much a product of per capita economic growth as population growth per se. Indeed, as Gardiner (2002) points out, it is wrong to think that more is necessarily better than less when it comes to the number of one's offspring. Be that as it may, the model is the same. Absent barriers to entry, public goods will be overused by individuals acting out of their self-interest, even as the (total) cost of doing so outweighs the (total) benefit. Short of reducing the population, Hardin canvasses the alternative of privatizing these resources to the extent that is possible. Where it is not, he is unimpressed by the power of conscience, and instead embraces the idea of coercion as a means of regulating our behavior, including limiting the choice to procreate without restriction. In this sense, Hardin's article is much less pessimistic than the uses to which it has been put. The Tragedy of the Commons does not drive to the conclusion that we are doomed, but rather that, absent coercion, we are doomed. But in its own way, Hardin's essay is also perhaps too optimistic. For what it does not address is how, and why, we should choose to subject ourselves to such coercion in the first place.

Any formalism maybe correct but nonetheless not be instanced. So you can agree with the logic of the Tragedy of the Commons without thinking there are any real-world cases of it. But that is too quick. Just as in the case of the axioms of rationality, here too the formalism involves a compromise between realism and mathematical tractability. And here too we need to ask whether it helps or hurts, whether the price we pay for getting the second over the first leaves us with enlightening results. Here the enlightenment we seek is how pessimistic we should be about the prospects for agreements to limit greenhouse gasses.

How much should we heed the pessimism that the logic of the Commons prompts? One reason not to embrace its pessimism is that in fact in some cases groups do in fact overcome its "logic." As Elinor Ostrom (1990) shows in *Governing the Commons*, such cases are sometimes about arrangements of extraordinary long-standing. Communal tenure in Toerbel, Switzerland, dates back as far as 1224 (Ostrom 1990, 62), while common lands in Japan date back to the Tokugawa period (1600–1867) (Ostrom 1990, 65). Moreover, as my colleague Bonnie McCay (1987) has shown, even fishermen can collectively regulate the commons under the right circumstances. But the rub is that what makes for the right circumstances is quite limited; too limited, one might think, to apply to the forces in play in climate change negotiations. As Ostrom argues the case, among other factors, what differentiates the cases that work from those that don't is that (graduated) sanctions are enforceable (which thereby create barriers to reentry). That is just what someone like Scott Barrett (2003) argues greenhouse gas reduction arrangements (like Kyoto) between nation states have lacked. At the most vulgar level, absent World Government, why would nation states not respect the logic of the commons? For they lack just what communities that have struggled to self-regulation have – enforceable sanctions.

II

The great virtue of game theory is that it sidesteps the iterated challenge of plumbing the choices of others. By restricting itself to situations in which pursuit of rational self-interest is all that matters, game theory offers solutions that obviate the need to "read" the minds of others and with it, the need to read their reading of others and so on. Thus, in a traditional Prisoner's Dilemma, all that matters is minimizing one's exposure to jail time. Prisoners A and B can each choose to cooperate with the police or remain silent, but they cannot communicate with each other. Then, if they are presented with the prospective jail time shown in Table 7.1, both will cooperate even though they would each do better if they both remained silent. The reason is that A does better by cooperating whether B cooperates or remains silent. And B does the same whether A cooperates or remains silent. This logic overrides any "gamesmanship," any need to try to fathom what he will do if I do x. In fact the Tragedy of the Commons has just this structure, it is just that the payoffs are in terms of productivity instead of years in jail and the choices are whether to exploit or show restraint. Suppose two people have access to a common lot. If they both show restraint, the land will yield a total of 4 bushels of grain, as it will if one of them shows restraint. But if they both try to exploit the land, its overall productivity decreases, as seen in Table 7.2.

TABLE 7.1 The Prisoner's Dilemma

	B cooperates	B remains silent
A cooperates	A:5 B:5	A:0 B:10
A remains silent	A:10 B:0	A:1 B:1

TABLE 7.2 The Tragedy of the Commons

	B exploits	B shows restraint
A exploits	A:1 B:1	A:4 B:0
A shows restraint	A:0 B:4	A:2 B:2

In the case of the commons, everyone wants as much as he or she can get and nobody wants to see the collapse of the commons. Climate change would seem to be identical: everyone wants to have as much carbon output as he or she can (absent cost comparable alternatives) and nobody wants to see (at least extreme) climate change. But is this characterization actually correct even when it comes to the case of the commons even when there exist no barriers to entry?

Consider a farmer who declines to put his cows to pasture on the commons. He knows his restraint will simply leave more for others. But still he acts. In one instance, imagine he is driven by a doctrinal commitment that demands "respect for Nature" irrespective of the material consequences that follow from obedience to that doctrine. Perhaps we will say, albeit reductively, that, in the terms of his preferences, obedience yields him "benefits" that outweigh the "costs." But whether we render such interests internal to the economic calculus or not does not really matter. All that matters is that, with such interests in hand, the clash of interests that drives the logic of the Commons is broken. The step from "if I show restraint, others will simply benefit and so there is no point in doing so" is replaced by the "if I show restraint, others will simply benefit and so be it." In doing so, the farmer steps outside the logic of the commons by embracing interests that it assumes are not in play.

Perhaps the farmer was not so high minded after all. Instead, he had his eye on a rich landowner whom he thought might hire him because he showed the appearance of good stewardship, even if he thought to himself that this was a mere gesture, doomed to failure by the actions of others. Here too we need not treat these interests as externalities. We can put numbers on the farmer's calculus. His actions can come out as rational based on those numbers.

But wait! Is the logic of the Commons undermined by any of this? Surely it does not demand that *everyone* march to its rhythms. As long as some people do, and their needs outstrip available resources, the problem will arise. One farmer stepping outside this logic does not change this. Nor do many. As long as there are remaining farmers with enough resources to buy more and more cows enough stay in place to create the problem, irrespective of the actions of others, and the logic of the commons still marches on.

Suppose most farmers follow the leader but some do not. But suppose the holdouts do *not* have enough resources to create the problem. They will show no restraint but their lack of restraint won't be enough to create the Hardin's outcome. Why might this be? A lemon tree stands on the village green. Villagers are asked to take no more than their fair share. Most do but some do not. If there is a market for lemons beyond the village then there will be an incentive for the holdouts to take all of the lemons. But if there is no market beyond the village and no market within the village (since everyone gets some lemons), satiation will limit the holdouts take. After all, how many lemons can one person make use of? This is a situation of stable partial cooperation. A parallel in climate would be if most countries restrained output but some did not, yet their unrestrained output was nonetheless restricted because of the size of their economies. There is, as it were, only so much

CO_2 that Monaco can produce given its geographical size, even if it changed its primary output from gambling to cement production.

That partial cooperation might be stable is surprising at first blush, even when the damage of noncompliance is limited. At the level of the day to day, it offends canons of fairness. If you won't show restraint, why should I! Of course one can witness the destructive downward spiral that this sentiment can lead to in daily life as much as in international affairs. Whether it is road rage or trade wars, when one party's lack of restraint is not matched by another, it can provoke the first to escalate their behavior and we are off to the races. What stops this happening in the first round, even in the face of some noncomplaint players, is that the benefit to those who choose to comply outweighs the costs, even when some get a free lunch.

Hume (1978, 538) writes that:

> Two neighbors may agree to drain a meadow which they possess in common; because 'tis easy for them to know each other's mind; and each must perceive, that the immediate consequence of his failing in his part, is, the abandoning of the whole project. But 'tis very difficult, and indeed impossible, that a thousand persons shou'd agree in any such action; it being difficult for them to convert so complicated a design, and still more difficult for them to execute it; while each seeks a pretext to free himself of the trouble and expense, and wou'd lay the whole burden on others.

But this problem of enforcement of an equal burden when the numbers are large does not necessarily doom the success of project. In fact, even if there are just two neighbors, what if one refuses to cooperate? Is the project necessarily doomed? Following Hampton (1987), consider the circumstances in which the benefit to one who goes it alone outweighs the costs. Then it is in his rational interest to go it alone even though his neighbor will be a free rider. (Notice though that this calculus may be different if these neighbors have other opportunities to share costs. Then acting alone and allowing free riding sets a precedent for more free riding.)

Of course each party may be tempted to free ride and then a coordination problem arises with a payoff preference ordering like that of Table 7.3 for A. That is to say, seeing the meadow drained is better than seeing it not drained, and getting it drained for free is an even better outcome. Of course the same is true for B, as seen in Table 7.4.

TABLE 7.3 A's Preference Ordering

	B cooperates	B free rides
A cooperates	2	3
A free rides	1	4

TABLE 7.4 B's Preference Ordering

	B cooperates	B free rides
A cooperates	2	1
A free rides	3	4

Under these circumstances, with a low number of players, muddling through to the second best outcome for both may seem like the likely outcome. Each may try to stare down the other, feigning indifference to draining the meadow. But more likely, sharing will triumph because of the surprising bias we seem to demonstrate toward fairness.

In a standard Ultimatum Game with two players, one of the players has all of the resources and makes a take it or leave it offer of a share of them to the second player. If the second player rejects the offer, neither gets any of the resources. Now in a one-off game, it would seem rational for the second player to accept any offer, it being better than the alternative of nothing at all. And hence it would seem rational for the first player to offer only a small portion of the whole. Yet empirical evidence does not support this view. Players in the second position tend to reject offers that deviate much from a fair share. And players in the first position tend to offer close to what is a fair share. Why should this be in a one-off game? The answer is that both players may have learned certain general "lessons of life" about how people behave. So the first player plays by a general rule that the second player is likely to reject an offer far off from a fair division. (Although Dan Ariely, 2010, reports that this generalization is not true of economists playing the Ultimatum Game!) Why might such lessons of life withstand the test of time? One reason is that while revenge (for the second player insulted by what was perceived as an "unfair" offer) may not be in his or her self-interest, the threat of revenge is. And to be credible, the threat of revenge has to backed up by (at least) some acts of revenge. Another more intriguing idea discussed by Airely (2010) is that revenge may be pleasurable. Ariely reports experiments in which a wronged player (that is one who judges he or she got an unfair offer) in an ultimatum game can seek vengeance, but only at a high price. In the setup for such an experiment, the second player can force his opponent to lose funds but has to pay $2 for every $1 his opponent is to lose. Not only do experiments demonstrate such behavior in which the wronged player willingly pays such a high price, but imaging of subjects in action suggests that brain loci involved in reward are activated. Such pleasure may be transient, only to be followed by regret about how things seemed in the moment. But if it is more long-lived than that, then acts based on revenge are perfectly compatible with considerations of rationality.

So two players meeting, even if only once, may have reliable reasons to play fair based on a general rule of life. Does this calculus change as the number of players increases? I think we all know the answer to this question. As the ratio of free riders to all players goes down, we are more willing to tolerate them. As the numbers rise, a "fair" share for each player gets smaller and smaller. At the same time, the impact of each player as a free rider diminishes. If you and I share a field and only I pay to drain it, because the benefits to me of doing so outweigh the costs, you get the same benefits for free. If there are 1,000 who share the field and 999 pay, you still get the same benefit as each who pays does, but you only get 1/1000 of the total benefit, unlike the first case in which case you got 1/2 of the total benefit. As such, as the ratio of free riders to the number of total payers goes down, the cost-benefit ratio to paying players becomes more favorable. Moreover, add to that the fact that if there is pleasure in vengeance, it too is (inversely) sensitive to this ratio. So not only will free riders be increasingly rational to tolerate as the ratio of free riders to the number of total payers goes down, but *willingness* to tolerate them will go up at the same time.

Of course the climate problem is not like Hume's field, and not like it in two respects. If we analogize reducing carbon output to draining the field, parties that choose to do the former have to be significant enough in their output for their restraint to make a difference. But second, in doing so, they need to be confident that any free riders won't use such constraint as an opportunity to increase their output. Let us consider these one at a time.

III

Suppose the United States and China decided to together restrain carbon output significantly without regard to what others chose to do. Together they constitute 42% of total output, so their restraint, if significant, would be significant in a way that the decision of smaller countries that decided to go it alone would not be. And as with Hume's field, such action might be rational for them to do, even if others were free riders, as long as the costs outweighed the benefits. At least it is if we can assume restraint in carbon output by them did not lead to an increase in carbon output by others. That is to say, if we can assume that the output of free riders would remain stable and thus lead to a partial cooperation stable outcome. If that were the case, the crucial issue is whether partial cooperation would be high enough to avoid the effects of climate change. Gardiner (2002, 412) argues that to avoid climate change almost all parties need to participate. Of course if all countries agreed but large emitters did not, we would be in trouble. But countries like Togo or Monaco could be hold outs with limited impact. It is certainly true that significant participation would be required for any of the parties to reap the benefits to make their own participation worthwhile, even assuming free riders would not increase their output in the face of the restraint shown by some. Why might free riders actually increase their output under such circumstances? The answer is of course a function of where you stand. We should not assume that the cost-benefit analysis is the same for all countries when it comes to the tradeoff between restraining carbon output and its costs, even for a free rider! Just as foregoing an increase in household income by $1,000 is wildly different if your current income is $1,000 as opposed to $100,000, so too, foregoing the economic growth by limiting carbon output is also a function of where you stand. And so is foregoing exploiting the opportunity for economic growth created by the restraint of others who value things differently.

Suppose countries A and B constitute 60% of global carbon output and countries C and D the remainder. A and B decide the benefits of cutting total carbon output in half is worth it, even if they have to do it all themselves. So, if C and D act as free riders and maintain their current output, A and B will have to cut theirs to one-fifth of global output. If C and D value the benefits of cutting carbon output the same as A and B (at least when it is free) then we have a stable outcome with partial cooperation. But this is not the only stable outcome. Assume that economic growth and carbon output are tied. A and B think economic growth is not worth it without cutting total carbon output by 50%. But suppose that calculus is different for C and D. Suppose they think the value of economic growth outweighs the consequences of carbon output as long as it does not exceed current levels. Then they have an incentive to take up the slack in carbon output produced by A and B's restraint. This looks like the Tragedy of the Commons all over again. If C and D can be expected to act as outlined, A and B have no incentive to constrain their output in the first place. So the original levels of output are stable. Will there be other stable levels between the status quo and a 50% cut as a function of C and D's valuation of the tradeoff between growth and the consequences of carbon output at any particular level? Suppose A and B want to see a 50% cut, while C and D only value a 25% cut. Then if A and B act to cut total output by 50% there is an opportunity for C and D to increase their carbon output by half of A and B's cut backs. But then A and B will be "paying" for a cut back of 50%, which seemed worth the price, only to end up with a 25% cutback. As such, A and B only have an incentive to cut back to whatever level that C and D value (as free riders).

Notice how on this analysis then, absent internationally enforceable agreements, the free riders dictate the level of constraint unless their maximum output is too small a percentage of total output to make a difference. So the challenge of cooperation comes in two different grades. The most obvious is, if I constrain output, will you? But the second is, if I constrain output, will you at least not increase your output? Securing cooperation for the second of these may seem much easier than the first, but even for the second, finding a way forward turns out to be surprisingly difficult unless all the parties can find a way to harmonize their goals.

That said, on either variant what we have here is an instance of a global good that depends on the coordinated efforts of many parties. The question is whether it is possible to win such cooperation in an international setting that lacks the kind of readymade enforcement mechanisms that a state has. To see that it is we need only turn to Scott Barrett's (2007) analysis of the Montreal Protocol, which is the most salient example of a climate treaty that worked, and examine how and why cooperation was achieved and the effects of free riders blunted.

The Montreal Protocol (1989) was negotiated in response to broad concern about stratospheric ozone depletion caused by chlorofluorocarbons. When chlorofluorocarbons (which were in widespread use as a refrigerant from the 1930s on) are released into the atmosphere, they release chlorine in the stratosphere as a result of exposure to ultraviolet radiation. Chlorine (Cl) reacts with Ozone (O_3) to produce oxygen and chlorine monoxide (ClO) which then combines with another ozone molecule producing chlorine and oxygen: $ClO + O_3 \rightarrow Cl + 2O_2$ thus allowing the cycle to begin again. The resulting depletion of the ozone layer allows more ultraviolet radiation to reach the earth, which not only increases risk of skin cancer in humans, but also has damaging effects on both marine animal life and crop production.

Avoiding ozone depletion is like avoiding carbon output in that both extract a price. But there is a difference: a plentiful supply of less damaging alternatives to chlorofluorocarbons exists even though they are more expensive. On the other hand, as I have argued, that is not the case when it comes to replacing fossil fuels unless we are willing to reduce or postpone the anticipated rate of growth in energy demand associated with anticipated economic growth in the Developing World.

The Montreal Protocol stands out as an agreement that worked in a way that climate negotiations have not worked. But the difference hinges on much more than the ready supply of (albeit more expensive) alternatives to chlorofluorocarbons. Barrett (2007, 77–83) argues that a combination of incentives and penalties was crucial to success along with a favorable cost-benefit ratio. When it comes to incentives, rich countries offered to pay for the differential costs of alternatives to chlorofluorocarbons for poor countries. But just as significantly, hold outs (rich or poor) suffered the threat of trade restrictions imposed by countries that signed on as parties to the agreement. (The trade restrictions here were both on chlorofluorocarbons and products containing them.) Now what is interesting about these restrictions is that they only have their intended effect if the ratio of those signing on to hold outs is high enough, which is why the treaty only took effect conditional on a minimum number of parties signing on to it. For with an unfavorable ratio, those not signing on would have only a limited restriction on the size of markets with which they could trade. Moreover, under those circumstances, those signing on would be disadvantaged in such markets because of their use of more expensive alternatives to chlorofluorocarbons.

What is important to notice here, and significant about the Montreal Protocol, is that, even absent the readymade enforcement mechanisms of the state, it functioned to undermine free riders by not coming into effect unless enough parties were on board, thus rendering the ride of those free riders no longer free.

Now, in theory, there is nothing to stop the implementation of exactly this strategy for CO_2 emissions. An agreement, made to put a price on carbon, could be conditional on enough parties agreeing, with trade restrictions then imposed on any hold outs. So what prevents theory from becoming practice? Barrett (2007, 94–95) argues the answer lies in the comparative cost-benefit ratios when you compare ozone restoration and climate protection (see Table 7.5). But this contrast is based on climate calculations by Nordhaus and Boyer (2000). As Barrett points out, incorporating the Stern report's (Stern 2007) very different discount rate (of 1.4% as compared to Nordhaus' use of 3%) radically transforms this estimate, producing a comparable ratio to the ozone calculus of 10:1. The record of inaction on

TABLE 7.5 Cost-Benefit Ratio for Ozone versus Climate Protection (Barrett 2007, 95, Table 3.3)

Goal	Cost*	Benefit*	Ratio
Ozone	$250	$2,775	11:1
Climate	$1,365	$681	0.5:1

*in billions of dollars

climate as compared to ozone suggests that, if the Stern calculation is correct, then the cost-benefit calculus is not in fact what is at issue.

But this glosses over a crucial issue. Differing judgments about the discount rate are a way of representing the relative value of the tradeoff between the present and the future that take into account a number of different factors; including the rate of return on capital and the rate of technological innovation. But as we saw earlier, the one thing the varied valuations of the discount rate don't put in question is a judgment of how much to care about the future as opposed to the present. Yet this is perhaps the crucial variable in play if we want to understand *political* judgments being made about climate policy today. Moreover, when comparing ozone policy to climate policy, the effect of this judgment is very different because although the costs of both ozone and climate policy may fall on those alive now, unlike climate policy, significant benefits of ozone policy are felt in the present as well. (That is to say, those who pay the costs of limiting ozone now are themselves the beneficiaries of avoiding an elevated risk of skin cancer.)

Like it or not, let us assume that in addition to the kinds of factors that go into an economist's derivation of a discount rate, valuing the lives of future generations per se differently from our lives *does* figure when it comes to political judgments for better or worse. Such a discount is not what drives the disagreement between Stern and Nordhaus. But, whatever the basis of their disagreement may be, if we add in such a multigenerational valuational discount, then the difference between their discount rates will get swamped out. And if the further we go into the future the greater that valuational discount, given that the benefits of avoiding the risk of severe climate change will occur far out in the future, we will be left with an unfavorable cost benefit ratio when it comes to acting to avoid climate change. And if that is true for us relative to generations that will follow us, then it is also true any future generation relative to generations that will follow it. This is a central element of what Gardiner (2011) calls "the intergenerational perfect moral storm."

IV

That said, let us assume for the purposes of argument, that notwithstanding these problems, that a number of players got the ball rolling. Notice, to do so, à la Montreal Protocol, they do not have to be large in number, or large in total carbon output, but rather large as export markets for the rest of the world, so that their threat to ban imports from countries that don't join them is economically meaningful. In fact, as we will see in the next chapter, the U.S. Waxman-Markey climate bill tried to achieve something like this.

Ignore for the purposes of discussion the issue of whether such a regime would pass muster with international agreements that ban trade barriers. (The political reality is most likely that whether it would or not would depends on the number and size of the players trying to impose such a regime on others.) A generally unexamined issue is how enforceable compliance with such agreements would be. Indeed, despite its formal success, enforcing the provisions of the Montreal Protocol on the use of chlorofluorocarbons has proved challenging because of the widespread use of residential air conditioners that are serviced by small companies. (See UNEP 2007.) One can only imagine a parallel set of problems of compliance where it to come to trying to regulating a carbon regime.

But Barrett (2007, 158–164) argues that under some circumstances, technology offers a readymade solution to this challenge that obviates the need for direct enforcement. Historically, the major source of coastal oil pollution near ports was because tankers did not segregate their oil cargo from their use of water for ballast. Thus cargo ships would regularly fill empty oil tanks with sea water to adjust their ballast only to then release the now oil contaminated ballast water close to shore prior to taking on new cargo.

Attempts to enforce prohibitions on this by monitoring ballast practices are practically impossible given the size of the oil cargo fleet and the geographical dispersal of that fleet. But in 1972 the United States mandated a domestic technical design standard for tankers that segregated water and oil. The law was designed to encourage international adoption of the same standard by banning nonconforming vessels from United States by 1976. By 1978 an international agreement was signed that had the same features as the Montreal Protocol. Enough parties signed so that the effect of banning ships from nonsignatory countries from their ports drastically reduced the value of such ships. Now an agreement with the same features to punish nonsignatories could have been signed to simply ban the practice of ballast water release but the innovation of incorporating a technical standard into the agreement had the great virtue of making compliance easy and reliable. One inspection can establish that ballast and oil tanks of a vessel are segregated and hence ongoing oversight to ensure oil contaminated ballast is not discharged close to shore becomes unneeded.

Might such a technically based solution be feasible when it comes to limiting carbon output? That is, a technical solution that is easy to verify and can be widely adopted, thus creating a disincentive for parties not to comply because of the fear of loss of trading privileges with the large number of confirming parties. If we focus on the major sources of carbon output (electrical generation and motor vehicles), in theory there is no reason such a technical solution might not be possible, in the sense that like tankers, one time inspection is all that is needed to determine conformity to agreed upon specifications. A coal plant can't masquerade as a nuclear plant and it is easy to distinguish an electric car from a gas powered vehicle.

But, as Barrett (2007, 162–164) points out, the widespread adoption of uniform technical standards has a much greater chance of success when they are driven by two other features beyond the threat of trade restrictions on nonconforming parties. The first is a matter of economies of scale. While it may be cheaper to produce a tanker without segregated ballast and oil tanks, building one kind of tanker with one kind of design may be cheaper in the long run. The second is a matter of externalities. A tanker only serving ports in one country need only be subject to the design specifications of that country. But a tanker that serves ports in many countries has to conform to the specifications of all of them. That creates an incentive for designs to conform to the standard of the most demanding country, assuming differing standards are compatible with each other.

Now these features are already true of some elements of motor vehicle production. Car manufacturers today increasingly use uniform "platforms" (that include common engines) internationally on which differing exterior designs are attached because of the economies of scale. And while there are still cross-regional differences in pollution and safety equipment standards, there is increasing uniformity within regions driven by the requirements of cross-border travel and sales. So there is reason to think these forces might also attach to the production of nonfossil-fuel vehicles, were they mandated in significant export markets for global car manufacturers. Not so, however, when it comes to power generation for two reasons. First, such plants are often highly subject to design features shaped by their specific location which trump considerations of uniformity of design. (Think only of hydroelectric dams.) And second, unlike tankers or cars, power plants don't travel across borders and thus across different regulatory regimes. (Although, if the power produced by such plants travels across such borders itself it offers an opportunity for some harmonization of intraregional standards).

But even though cars capture the characteristics in theory that Barrett argues maximize the chances for a technical specified standard to gain widespread adoption, as he points out (2007, 163–164), implementation is another matter. Quite aside from the absence of an infrastructure to support an alternative fuel for gasoline, what we also lack is a system to produce that alternative fuel (for example electricity) in the quantities needed that do not themselves rely on fossil fuels for the foreseeable future given the scale of demand.

Thus the prospects for the adoption and propagation of a technically specified standard to drive down carbon output are not good, at least for now, when it comes to both of the major sources of that output. But the fact of the matter is that, even if this was not the case, the much more telling hurdle that stands in the way of a Montreal Protocol model approach is that, unlike an ozone agreement, as we saw earlier, the cost-benefit ratio is not favorable for a large enough group of players to get the ball rolling with trade incentives. At least it is not given that those players lack a time horizon that is multigenerational and embrace a low enough discount rate when it comes to calculating the costs for future generations.

In the next chapter, I want to examine this hypothesis in the light of actual climate negotiations. Given that China and the United States together produce so much of the world's carbon output and together constitute a vast trade market, they are well positioned, if anyone is, to implement a Montreal Protocol strategy which would force others to follow because of the threat of trade sanctions. When we look at the sorry record of their failure to come to any agreement, is the cost-benefit calculus what is in play?

References

Ariely, Dan 2010. *The Upside of Irrationality: The Unexpected Benefits of Defying Logic at Work and at Home*, New York: HarperCollins. Kindle Edition.

Barrett, Scott 2003. *Environment and Statecraft*, Oxford: Oxford University Press.

Barrett, Scott 2007. *Why Cooperate?*, Oxford: Oxford University Press.

Gardiner, Stephen 2002. "The Real Tragedy of the Commons," *Philosophy and Public Affairs*, 30: 387–416.

Gardiner, Stephen 2011. *A Perfect Moral Storm: The Ethical Tragedy of Climate Change*, New York: Oxford University Press.

Hampton, Jean 1987. "Free-Rider Problems in the Production of Collective Goods," *Economics and Philosophy*, 3: 245–273.

Hardin, Garrett 1968. "The Tragedy of the Commons," *Science*, 162: 1243–1248.

Hume, David 1978. *A Treatise of Human Nature*, Oxford: Oxford University Press.

McCay, Bonnie 1987. "The Culture of the Commoners: Historical Observations on Old World and East Coast U.S. Fisheries," in B. J. McCay and J. M. Acheson (eds.), *The Question of the Commons*, Tucson: University of Arizona Press, 195–216.

Nordhaus, William and Joseph Boyer 2000. *Warming the World: Economic Models of Global Warming*, Cambridge: MIT Press.

Ostrom, Elinor 1990. *Governing the Commons*, New York: Cambridge University Press.

Stern, Nicholas 2007. *The Economics of Climate Change: The Stern Review*, Cambridge: Cambridge University Press.

UNEP 2007. *Illegal Trade in Ozone Depleting Substances*, www.mea-ren.org/files/publications/Illegal%20Trade%20in%20ODS.pdf, accessed July 8th, 2013.

8

NEGOTIATIONS GONE BAD

Embracing the twin goals of health care reform and climate change legislation in 2008, Barack Obama declared "We will be able to look back and tell our children that this was the moment when we began to provide care for the sick," and the moment "[w]hen the rise of the oceans began to slow and our planet began to heal" (Lizza 2010). So when he won the presidency, hopes were high that Obama would lead a legislative effort on climate legislation, and all the more so for having declared it a priority on grounds of national security. Moreover, if there was any question of whether he could achieve both health reform and climate legislation simultaneously, his ordering of the priorities was clear after he tied combating climate change to an overall energy strategy and declared that "Energy we have to deal with today. . . . Health care is priority No. 2" (Lizza 2010). So after taking office, it came as a surprise when those priorities were not only reversed but climate change was deemed a losing legislative venture by Obama's advisors who argued that the legislation, if it succeeded, would only cost him with the electorate unlike health care legislation.

But despite the White House's stance, legislation on climate began in the House of Representatives, which was under the control of Democrats at the time. Introduced by Henry Waxman and Edward Markey, HR 2454, The American Clean Energy and Security Act of 2009 was approved by the House of Representatives on June 26, 2009 by a vote of 219–212. At first glance, the Waxman-Markey bill was noteworthy for taking a unilateral approach to dealing with climate change. It thus stood in stark contrast to the sentiments expressed by George Bush on March 13, 2001, when he declared that he "oppose[d] the Kyoto Protocol because it exempts 80% of the world, including major population centers such as China and India" (White House 2001). Even if unilateralism seemed of limited value, given that the total output of greenhouse gasses attributable to the United States at the time the bill was drawn up was 17.9%, in fact the approach used a stick very much like the Montreal approach that exploited the U.S. economy's appetite for imported goods. Thus while the bill set unilateral goals for U.S. reductions of 17% in greenhouse gases over 2005 levels by 2020 and 83% by 2050, it also empowered the president to set "border adjustments" for "covered goods." These provisions were designed to increase the costs of imported goods manufactured by economies with no equivalent greenhouse gas restrictions to those being proposed in the bill.

Just how effective these sticks would have been is something we will never know since the bill was defeated in the Senate. But it is worth noting that while its provisions might have protected United States competitiveness by leveling the playing field with manufacturers of imported goods, it still might not have set in motion movement in the direction of the adoption of uniform standards discussed by Barrett in the last chapter for the following reason. The border adjustment mechanism was to be realized in terms of requiring manufacturers of noncompliant covered goods to purchase emission allowances. Now while that increases the cost for any imported item to the United States, it provides no incentive for the manufacturer to produce those goods with lower greenhouse gas emissions for its domestic economy or other export markets. Indeed the emission allowance mechanism does the exact opposite, since it allows the manufacturer to proceed with business as usual, instead of forcing a change in practice governing *all* units produced, whatever market they are to be sold in, in the way that we saw the imposition of technical standards can do. (At least it does, Christopher Preston pointed out, unless most of the market for the manufactured good is for export to the United States.)

Why did the legislation fail in the Senate? What did the House see in the bill that the Senate did not see? Putting the question this way is to implicitly assume a certain set of reasons to be for or against the legislation that are content driven. But while there is much to debate about in the content of the Waxman-Markey bill, the key factors that led to its failure in the Senate bear no relation to such internal considerations. Far from it, what a narrative of the legislation in the Senate reveals is a familiar tale of how legislation can succeed or fail for reasons quite unrelated to the content of the legislation. Those reasons may be rational, but what drives those considerations of rationality depend more often than not on the self-interested assessments of individual legislators. And as Lizza's (2010) sad tale of climate legislation's failure in the Senate shows, what counts as relevant to the calculus of perceived self-interest is often much broader and idiosyncratic than the content of the legislation itself.

Senate legislation on climate was initiated by Barbara Boxer of California who chaired the Senate Environment and Public Works Committee. However, since legislation in the Senate would require Republican support to be filibuster proof, because not all Democrats were going to support it, her efforts were quickly overtaken by legislation drawn up by a coalition of three senators that included Republican Lindsey Graham of South Carolina. Graham, along with John Kerry and Joseph Lieberman, had reasons to want to move "big" legislation. Both Graham and Kerry had not distinguished themselves in doing so before and were anxious to bolster their respective reputations. Meanwhile Lieberman was anxious to make peace with the Democrats after having been reelected in 2006 as an independent following his loss of the party primary for the seat.

With the Democrats short of votes because of opposition from senators in their party who represented coal producing states among others, the bill's managers knew they needed at least four Republicans to join Graham in support of the bill and targeted Susan Collins, Olympia Snowe, Scott Brown, and George LeMieux as prospects to make up a filibuster proof majority of 60. How likely this was to have happened is an object of political speculation. The White House was certainly dubious. But before this question could be put to the test, Lindsay Graham withdrew support for the bill. What caused his change of heart? The minority view is that he came to realize that he had underestimated the political price he would have to pay among Republicans and simply wanted a way to back out. On this view a decision by the Senate majority leader (Harry Reid) to bring an immigration bill to the floor before the climate bill provided a convenient excuse. "Immigration was interjected before we rolled out the [climate and energy] bill not because anybody's serious about passing it, but because Harry has got a political problem with the Hispanic community" complained Graham. "It makes the heavy lift of energy and climate impossible and everybody knows that" (Beutler 2010).

But the problem with this view is that a much more graceful way to achieve his goal would have been for Graham to acquiesce to immigration going first and then work to tie up Senate time on it thereby precluding time for the climate bill that he no longer wanted to support. (Indeed, as an author of the climate bill he had all the tools at his disposal to delay it moving to the floor by demanding changes in its content that he could choose to be hard for other supporters to swallow.)

The alternative (majority) view is that, in fact, Graham actually did not want to see immigration legislation move and was willing to do anything to stop it, even though he knew it had no chance of passage. In fact, Graham was so willing to do anything to stop it that he even issued a threat to filibuster his own climate bill if it actually came first simply to prevent immigration coming the floor at all. Just why he did not want to see immigration legislation move comes in two variants. One was a desire not to have the Republican Party come out against it for fear of alienating Hispanic voters. The other was more specifically tied to his friend John McCain's reelection bid in Arizona. McCain had already been forced to tack right in supporting Arizona's own restrictive legislation on immigrants and the last thing he wanted to be forced to do was to vote on federal legislation. However, Harry Reid had his own reasons for wanting to force a vote on the issue. He was in a tight race for reelection in Nevada and anxious to win favor with Hispanic voters in that state.

With Graham's ultimatum in place, the route to save the climate bill went through the White House which was asked to pressure Reid into withdrawing the immigration legislation. But this the White House declined to do. That is to say, Obama's advisors decided to allow Reid's interests (and with it their party's) to trump moving climate legislation.

Now from the point of view of party interests, this choice may have been the right one. On that calculus, loss of the Senate majority leader would have been a serious blow. Certainly the choice would seem to be right for Reid to proceed with immigration whatever the White House position. And finally, given that Graham's reasons for supporting the climate bill in the first place were not related to its content as much as an interest in passing "significant" legislation, sacrificing it for the welfare of Republicans may well have been the right decision for him as well.

Of course the "rightness" of these decisions is relative to narrow and parochial interests. But that is the point of my telling of this sorry story. The challenge of trying to understand the rational actions of "players" in formalizations of decision making always depends on a proper characterization of how they see their interests. But here there is another layer of challenge as well. If we begin with nation states as the players it makes no sense to ask about how they see their interests except in a metaphorical sense. It is a matter of how those acting on their behalf see things. But it is naïve to think that political actors simply act on behalf of their national interests even when they are charged with acting on behalf of those "national interests." Instead the interests that are relevant occur on many different levels – from the perceived national, to the political parties, to the individual. And capturing this chaotic mix of motivations along with attributing the right weightings to each of them is a tremendous challenge even before embarking on any attempt to formalize the process. Nor is it clear what the payoff of formalizing it in the first place in this context, as we have seen in other contexts, given the amount of simplification that doing so demands.

Moreover the price we pay for achieving such formalizations is not just simplification but also the assumption of a degree of determinism that underweights the role of chance and luck. For it seems to me that what the demise of climate legislation in the United States Senate shows is in large part the role of luck: both good and bad (unless the minority view of Graham's motivation is correct). Good luck in that Kerry, Lieberman, and Graham were looking for a "big" piece of legislation to bolster their reputations when the opportunity for a climate bill came in to view. Bad luck in that the Tea Party rose to prominence when it did, and that that resulted in a serious challenge to Harry Reid when it did. Can we make such assertions with much confidence? But for Kerry et al.'s political needs, the bill still might have moved forward. And but for Reid's challenge, the bill might still have floundered.

II

Historical events can come with varying degrees of resilience to perturbation of their causal histories. AIDS spread from Africa to the Western gay community around the world by a particular causal chain. Purportedly Gaëtan Duga, who worked for Air Canada as a flight attendant, initiated that chain. Assume he did for the purposes of argument. But even if he did, a chain of events very much like the one that actually occurred would have occurred, initiated by another "patient zero," absent his role. The outcome here is thus quite stable. Contrast that to the impeachment trial of Bill Clinton. His affair with Monica Lewinsky lies in the causal history of that event. But for the affair the impeachment trial would not have taken place because Clinton's political opponents had no other cause in hand that could have served as an alternative cause of action. In this case, unlike the AIDS case, the outcome is quite "iffy" in its dependence on the actual causal history. (For a formal treatment of these ideas see Ben-Menahem 1997.) In talking about luck, what we are really talking about are cases of such iffy causal histories. In this sense, if the minority view of Graham's motivations is correct, luck is not relevant. If not for the immigration bill he could have, and likely would have, found a different reason to scuttle the prospects for the climate bill. On the other hand, if the majority view is correct, just the opposite is the case. Graham's decision to withdraw support is not stable under different causal histories since the immigration bill was his reason for withdrawing.

But what if we look at this causal history with a much lower level of magnification? The Waxman-Markey bill was an attempt to reach global climate regulation by one particular route that depended on unilateral action by the United States that failed. But if David Victor (2011) is right, that failure itself is stable under perturbation of its causes, and many different routes would have led to the same outcome. Or to put it differently, even if Waxman-Markey had become law it would not have succeeded in leveraging a global agreement. Why?

For Victor, what sets the Montreal Protocol apart and makes it different from attempts to regulate climate is that the Montreal Protocol worked for a reason that most people ignore – it included escape clauses for uses of chlorofluorocarbons that were deemed essential. By Victor's lights that is crucial because it has the effect of creating a weaker regime than one dictated by set caps and timetables. Why is that a virtue rather than a vice? Because, Victor argues, without it, no climate agreements can ever be expected to be reached, let alone implemented. The reason he argues is that "when governments don't know what they can achieve in advance a large measure of flexibility will be needed" (Victor 2011, 224). But if agreement was possible for chlorofluorocarbons why not for carbon? If escape clauses worked with a binding regime for chlorofluorocarbons then why would they not work for carbon? Although Victor does not say it explicitly, I think the answer is simply that the scale of uncertainty in the case of carbon is much larger than it is for chlorofluorocarbons and it is so for three reasons: first, the amount of the economy touched by carbon is immensely larger than that touched by the use of chlorofluorocarbons. Second, there were ready alternatives to chlorofluorocarbons, which, even if more expensive, could be quickly deployed "at scale" on a timetable as needed by the dictates of the economy which is not the case when it comes to alternatives to carbon. And third, as a consequence of these first two reasons, the scale of uncertainty in the economic costs of adopting binding caps and timetables is thus immeasurably larger for carbon as opposed to chlorofluorocarbons. These three reasons combine to make the needed escape clauses for climate agreements not something specific as they were in the Montreal Agreement (D'Souza 1995) but themselves subject to uncertainty.

III

Waxman-Markey was an exercise in the unilateralism designed to pull other nations along with it through border adjustments and, of course, it was China that would have been most affected by those adjustments given the size of its exports to the United States. But China and the United States had been involved in long standing, if intermittent, negotiations about climate that began before Waxman-Markey and still continue, in both multinational and bilateral settings. To say the results of these efforts have been meager is perhaps an overstatement, notwithstanding Durban climate agreement. For although China and the United States were parties to the international agreement in Durban in 2011 to "as early as possible but no later than 2015" to adopt a "legal instrument or an agreed outcome with legal force . . . to come into effect and be implemented from 2020" (United Nations 2011, 2), it is important to notice that this language leaves the content of such an instrument or agreement undefined, and thus could be satisfied by language about carbon *intensity* as opposed to *absolute levels* of carbon use.

How should we see the failure to come to an agreement of any more substance than this to date? Is that failure an outcome that could easily have been different given different circumstances? Or is the outcome stable under perturbation of its causal history so that not much would make a difference in the outcome? To examine this question I want to canvass theories of just why China-United States climate negotiations have been so unproductive. Such theories are both numerous and variegated. In the first instance at least, I am going to be less interested in the likelihood of them being determinative as much as what the consequences would be if they were. For some of these theories point to causes of the failure to achieve agreement in which that outcome seem to be quite unstable. As such they hold out much more hope for eventual success as compared to other theories that lack this feature.

When the United States chief climate negotiator, Todd Stern, invited Vice Minister Xie Zhenhua to see a game involving Stern's beloved Chicago Cubs, reporters took note. That Stern would make the offer, and that Xie would accept, indicated a flowering personal relationship between the two that could only help the prospects for climate negotiations between the two countries, so the reporters argued. Perhaps. But of all the reasons that might explain why United States-Chinese negotiations have been so unproductive, the idea that the absence to date of any personal chemistry between the chief negotiators is crucial seems perhaps the least plausible. It is a nice thought of course. A night on the town and difficulties, if not solved, are at least eased. But chief negotiators don't work in isolation. Most of the detailed work is done by legions of "sherpas" who report up the line to them and, despite their title of "chief," they report up the line as well. Moreover, as experienced negotiators, Stern and Su could be expected to be much more disciplined than lay negotiators, or even politicians, and thus less likely to be affected by the personal "chemistry" between them.

But if personal chemistry is not likely to make a difference, a host (10!) of alternative theories of what has gone wrong abounds among both low-level negotiators as well as observers of the negotiations in both Washington and Beijing. Let us canvass them one by one in order of increasing plausibility.

The very conception of encounters between negotiators from the West and the "Orient" invites the expectation of the potential for a clash between two very different cultures. It is fear of causing unintended offence that dooms negotiations that creates a market for numerous books on etiquette for Westerners doing business in China. "Colors play an important part in Chinese superstitions" writes Stefan Verstappen (2008) variously emphasizing the importance of avoiding bright colors altogether, white (the color of death), any colors in presentations (in case they are "unlucky"), using gold for business cards (which conveys prosperity), and finally wrapping presents in red (which is lucky) and not blue (which is not). On Verstappen's account, color is not the only minefield for Westerners to navigate. Fortuitous times for meetings are only between April and June, and September and October. Not observing the correct protocol for entering a meeting with the senior person of a delegation in the lead can result in confusion. Not presenting a business card in the right way and studying your counterpart's card before slipping it in your pocket can cause offense. And flipping over a fish in a restaurant may be seen as causing bad luck, causing "the fishing boat that caught your fish to capsize" (Verstappen 2008, Kindle Location 1010).

Of course all of this mysteriousness runs along a two way street, as evidenced by Morrison and Conaway (2006, 550–554). Foreigners are warned that in the United States:

> Taking calls while others are in the room . . . is common practice.
> People may write on your [business] card. . . . This is not meant to show disrespect.
> When giving an item to another person, one may toss it . . .
> When not working . . . [y]ou may see people wearing torn clothing.

But when it comes to climate negotiations, the teams are staffed all the way up and down the line with cosmopolitan experienced negotiators quite unlike more insular and provincial parties (on either side) that may come to the table in business negotiations.

The question of global cultural difference notwithstanding, by all accounts, the United States and China do approach negotiating with differing cultural styles. The United States' side is seen as more likely to outline areas of potential flexibility and the range of that potential flexibility at the outset as a tactic, while the Chinese side is more prone to declare an unwavering bottom line at the outset and hold to it, if only as a tactic.

Another difference often noted between the two sides revolves around differing styles of leadership and autonomy. Except for the very top leadership, Chinese negotiators are often characterized, by both their United States counterparts and independent observers, as lacking autonomy. As a result, negotiations proceed (when they do at all) haltingly because of the frequent need to report to superiors and get new instructions. (Observers at Copenhagen described Su Wei and his colleges as increasingly exhausted because of the demands of long consultations with Beijing which took place at night because of the time difference between China and Denmark.) Moreover the way in which decisions are made (by committee) also affects top leadership who are wary about acting alone, even if they are technically empowered to do so in international settings. When President Obama took over the United States reins with his counterparts, Chinese Premier Wen Jiabao did not attend the pivotal meeting of "principals" because he was said to have been offended by not have been personally invited and sent a deputy, He Yafei. But in other encounters in Copenhagen, he was described as at a disadvantage because of their freewheeling and informal nature.

Further down the hierarchy of Chinese leadership another problem arises. The popular conception of Chinese government decision making as "centralized" belies the fact that it is also extremely balkanized. So while, in the end, all units may have to report to the Standing Committee, the number of separate units doing so for any one issue is extremely broad. The result is that in any area, including climate, reporting has to take place up parallel chains of command through many sectors of the government, each of which may have competing and conflicting interests.

The idea that the United States and China might act bilaterally on climate is embedded in the context in which they are seen as world leaders, setting a precedent for the rest of the world. In this sense, the context in which bilateralism operates is global. Yet many observers of the negotiations in Beijing have argued that this goes against current Chinese thinking, which is focused on domestic and regional issues. The political demands of managing China's high rate of urbanization, and the transformation of its economy, are considered all absorbing. Moreover, a focus on regional issues trumps a focus on global issues because both East Asian land and sea transportation are crucial to China's continued economic growth. That is not to say that China has no global concerns, but these are largely restricted to controlling raw materials and the energy needed for economic inputs by direct ownership, partnerships in foreign countries, or long-term contracts.

There is an idea among observers in Washington that China is a global power and has a duty to act as such. This is sometimes asserted as if it were a moral doctrine of self-sacrifice, suggesting that other global powers (including the United States) have acted as such for altruistic reasons which China is seen as selfish for not embracing. Despite the self-serving and naïve elements of this point of view, it is true that observers in Beijing are quite clear in reflecting the government view that China does not see itself as being ready to lead globally, and they hold to this view by asserting a particular view of what leadership entails. "Leadership," so the view holds, entails sacrifice. To lead on climate change policy requires a willingness to lead by example. Not only does China not feel ready to make such sacrifice as the price for leadership, but more tellingly, it sees the United States as failing to fulfill its duty as a leader by its unwillingness to lead by sacrifice as well.

Even if, on this view, China is seen as not "ready" to assume global leadership, a popular view propagated by Chinese elites is that the United States is trying to do everything it can to thwart the rise of China as a global power. Whether this view is asserted as a matter of conviction, or solely as a basis that plays well with "the masses," is a matter of dispute. Some observers in Washington think that it is taken more seriously by the Chinese leadership than observers in Beijing do. But on either interpretation, the claim made is that United States action fits into a broad historical theory of the behavior of "declining" powers and how they try to prevent the ascent of "rising" powers, purportedly based on the study of Germany in the late 19th and 20th centuries. With or without this historical grounding, one thing this view underscores is a stance on climate negotiations which sees them as part of a much larger set of negotiations that governs all aspects of the United States-China relationship. And even if the declining-rising power trope is largely for public consumption, nonetheless Chinese elites do take seriously the idea that the United States should be understood as doing its best to set up roadblocks for China on all fronts. This has the effect of framing the dynamics of all climate negotiations (and especially concessions) in a much broader context so that the climate negotiators do not even have limited control or authority over their own portfolio per se. Of course this is not to say that global features of the United States-Chinese relationship don't also get implicated for the United States side as well in climate negotiations. But all the same, in this case, it is less driven by the belief that any Chinese positions in one area are driven by considerations in all other areas that govern the relationship.

The seven considerations raised so far as explanations for why negotiations between China and the United States have proved so difficult are noteworthy in that they all revolve around what might be termed "standing" conditions. That is, features of the relationship, like negotiating style or leadership style, that can be expected to remain in place even though they may change at the level of detail or specifics. Thus even if a particular causal history were not to hold, it would be replaced by one very much like it and thus could be expected to cause the same outcome. As such, they are just the kind of considerations that lack the kind of "iffyness" discussed earlier and on that basis, if they explain the failure of negotiations, we ought to be pessimistic about the prospects for change. But that said, although these considerations have some degree of support by observers on both sides, three other aspects of the negotiations are treated as being much more plausible as critical explanations of what has gone wrong.

The recent use of the acronym BRICS has come to refer to more than just the initial set of countries that gave rise to the term (Brazil, Russia, India, China and South Africa). The term now also encompasses other countries like Mexico and South Korea. For some people however, these should not just be understood as a group of fast developing countries. Instead the idea of the BRICS should be seen as a challenge to the taxonomy of the Developed-Developing World which should be amended to recognize a third distinct category between the two. But in negotiations on climate between the United States and China, the old taxonomy has prevailed along with a pointed disagreement between the parties as to where China "fits." Neither side thinks that China is a developed country in all respects, but the United States position has been to argue that it is, at least in some respects and, as such, it should assume responsibilities appropriate to that station. On the other hand, the Chinese side has held fast to the idea that it is to be thought of as a developing country. This is not just a matter of opportunism on its part, qua its desire to be seen as a leader of the poor countries of the world, but is also reflective of its domestic priority of promoting development to raise annual rural income, which currently stands at $1,300 per person in comparison to $4,000 per person for urban residents (Buckley 2013).

Notwithstanding these perspectival differences, each side in the negotiations also suffers from surprisingly unrealistic views about just how much control leaders on the other side have. Thus Chinese observers often fail to appreciate how limited United States presidential power is, especially when one or more houses of the Congress are not under the control of the President's party. Moreover, they also fail to appreciate how much United States policy is sensitive to perceived electoral "cost" of it by politicians. At the same time, there is a widespread perception in the United States that the Chinese economy is run by "command and control" so that changes in policy should be much easier than the "messier" process in United States. But this view is unrealistic in four different respects. First, as already noted, political power in China, at the higher levels, is itself relatively diffuse and subject to a consensual process that has to balance many conflicting political interests. Second, as widely noted, one of the ways the national leadership wins provincial political acquiescence is by ceding significant authority to provincial governments in setting their own priorities. Third, since the time of Deng Xiaoping, corporate decisions have been increasingly freed from government central planning by deregulation, including coal beginning in 1979 (Peng 2009). Finally, it is generally accepted that there is an implicit compact between the Communist Party and the general population in which the latter accepts rule by the former in exchange for continued delivery of economic growth.

Whatever degree of command and control either government may in fact have, any binding agreement between them will require a serious commitment to its implementation. As such, the worry may be that those commitments will outrun either side's ability to implement an agreement, quite aside from their interest in doing so. However one way to sidestep this dilemma is to only commit to limits or caps for carbon output that are above the anticipated levels needed given the expected rate of economic growth. Moreover, doing this, more importantly, allows the parties to avoid the choice between growth and restraint of carbon output! In the case of China this has given rise to three different stances. The optimists argue that China's annual growth rate will peak in the next 10 years due to a shift away from energy intensive manufacturing. That will allow for a reduced rate of growth of its energy needs, they argue, and allow ample room for that growth to be accom-modated by renewables along with a reduction in the current fossil fuel portfolio, by phasing out and replacing the dirtiest coal plants. Thus the optimists think there is no price to pay with caps to limit global warming to no more than 2 °C. On the other hand, the pessimists think that China is unlikely to peak in its fossil fuel use until the 2030s or even later. As such, for them, there is a real conflict between energy needs and carbon caps except insofar as those caps are set high enough, even if doing so entails countenancing a world with more than a 2 °C rise. Finally, the agnostics are skeptical about any projections, and are thus leery about making any commitments whatsoever. Agnosticism itself comes in two variants; one based on academic work suggesting a high degree of uncertainty when it comes to choosing between optimism and pessimism. The second variant is based on a more general-ized political distrust about academic projections *tout court*, which, not surprisingly, is the view held by the Chinese *political* leadership. Still, whatever the differences between optimists, pessimists, and agnostics, they are consistent with each other in the view that no caps should be accepted as binding except insofar as they involve no (economic) sacrifice.

The case of the United States is different in that its projected energy growth needs are much easier to have confidence in because its growth rate is slower than China's. That said, the question is whether or not, if push comes to shove, in the end the United States would support binding limits except insofar as they were above the carbon output levels associated with expected economic growth.

In this regard, it is worth noting the limited ambition, and likely limited bilat-eral impact with China, of President Obama's 2013 unilateral climate initiative to limit the use of coal without effective CO_2 flue capture in electrical generation. While it sets the same goal as the Waxman Markey bill for 2020 (a 17% reduction over 2005 levels), it is silent on the more important, and likely economically costly, 2050 targets of Waxman Markey (an 83% reduction over 2005 levels). Nor does it impose import tariffs on countries that don't match its restrictions on carbon use. Moreover the 2020 goal is modest in the light of the effects of the 2008 recession on energy demand. U.S greenhouse gas output in 2005 was 7,196 million metric tons of CO2 equivalent but fell by 9.3% to 6,702 in 2011. Whether the initiative can be implemented remains to be seen, since political and legal challenges can be expected. But even if it is implemented, because it is in the form of an executive order, its longevity will depend on ongoing presidential support, not just by Obama but subsequent presidents as well. That said, there is an underlying economic reality that makes limitations on the use of coal essentially cost free: "fracking" has dramat-ically increased the supply, and lowered the price, of natural gas in the United States, rendering coal plants uneconomical. That is why no new coal plants are being built and the accelerated retirement of existing plants has its own economic logic. But that said, it would be naïve to think that coal is going to stay in the ground. The United States coal industry is already undergoing the process of transformation into an export market. That speaks to the reality that absent cooperative action, United States policy in and of itself will be of limited value especially looking forward over the next 20 years as energy demand grows dramatically around the world, including of course China.

The three reasons for the failure of United States-China negotiations just canvassed are deemed more plausible by observers on both sides than the seven reasons previously offered. I argued earlier that those seven reasons, implausible or not, have in common the feature such that, if any of them were true, they would likely produce an outcome that is hard to change. What about the three reasons we have just been considering? Here the picture is mixed. One of the reasons just canvassed implicates structural features of both sides; namely, the limited power of each leadership. Another of them (the degree of command and control) also reflects structural features of the Chinese economy. Such structural features, by their very nature are not easy to change and so, if failure is due to either of them, the expectation of an alternative (successful) outcome is doubtful. On the other hand, the third reason (the priority on development and the desire for carbon caps that don't require economic sacrifices) is not a structural feature but a matter of political preferences. As such, it is a better prospect for being upended. But how good the prospect is for this happening requires probing how firmly held those political preferences are. In this regard the picture is clear on the Chinese side since we know the outcome of debates that have taken place about trading a lowered economic growth rate for a lowered rate of increase in carbon output. On the other hand, in the case of the United States, for any politician to embrace a tradeoff in favor of restraining carbon output and giving up economic growth, the issue would have to be "off the table" as a contested area between competing parties. The fact that this has not happened makes it very unlikely that carbon caps would win support, beyond those envisioned in the modest 2020 goals of President Obama's plans, except insofar as they could be defended as consistent with ongoing economic growth.

In this sense there is a symmetry between the United States and China in that neither side is likely to be willing to countenance a lowered growth rate as the price of imposing limits on carbon output. In both cases the political risk of doing anything different is domestic and, as a consequence, it is a mistake to judge the failure of the two parties to come to an agreement as a failure of negotiations per se. Assuming domestic political considerations settle the matter, it is misleading to think that but for Chinese intransigence the United States would agree, or vice versa. The unwillingness of either side to enter into an agreement is not what is preventing the other side from doing so. To put this in terms of the discussion of David Victor's analysis, the key issue is not just uncertainty about the economic costs of adopting carbon caps. For that allows the impression that the costs of adoption would be acceptable if they were known ahead of time. On this analysis, things are even worse: what prevents agreement is the absence of a guarantee that there are no costs at all! But such a guarantee will be hard to provide. For, unlike chlorofluorocarbons, fossil fuels are so widely implicated throughout any modern economy that their knock-on effects on growth are hard to limit, either when more expensive inputs are substituted or, even more so, when total energy input itself is limited.

Of course, we might argue, were political leaders to take seriously the nonzero probability of the kind of worst case scenarios discussed in previous chapters, the idea that there is really a conflict between economic growth and limiting fossil fuels evaporates in the multigenerational long run. But absent some sort of conversion experience, in politics, it is the very short term that defines the terms of the debate, along with the high discount rate that is applied to future generations.

IV

During a trip to Beijing in the summer 2009, I only saw the sun for a few hours during the week I was there. If the atmosphere reminded me of the smog with which I grew up in London during the 1950s, everyone I spoke to assured me that it was fog not smog. Returning in 2013 for another week with even less sunshine, only one person still insisted the problem was fog and not smog. For what it is worth, he was a civil servant charged with developing educational programs on climate change for the Chinese public. The smog of 2009 was merely eerie, but the smog of 2013 was of a totally different order. It assaulted the senses with an acrid smell the moment the cabin door was opened on the plane on which I arrived. It was so thick and pervasive that one end of my hotel lobby looked murky from the other end. It was so irritating that I had to wear a mask to avoid coughing. And finally, it was so concentrated that it produced labored breathing.

High concentrations of particulate matter of 2.5 micrometers or less are considered particularly unhealthy because of their ability to penetrate deep into lungs. The World Health Organization designates concentrations of more than 25 micrograms of such particulate matter per cubic meter as unhealthy. So it gives a sense of how bad air quality has become in Beijing that the concentration of such particles in January of 2013 (during my second stay there) was measured at 700 micrograms per cubic meter.

Of course Beijing's smog problem is by no means unique as an instance of poor air quality in China today, although it is especially prone to long lasting smog events because it sits in a vast bowl and is subject to temperature inversions that trap both auto emissions and the flue gasses of its three coal plants.

While avoiding climate change primarily poses a tradeoff between costs to people now and benefits to those in the future, those who bear the costs of improving air quality (as in the case of reducing stratospheric ozone depletion) also reap the benefits of doing so themselves. That changes the calculus for political leaders of course, so it should not be surprising that the Chinese leadership has made improving air quality a priority. Now some have suggested that the drive to improve air quality can play an important role in avoiding climate change. The causes of poor air quality are themselves significant sources of greenhouse gasses other than CO_2, so one might think that restricting their output would kill two birds with one stone. However the effects of poor air quality on climate are unfortunately not so simple. "Smog" comes in two varieties; one is from the incomplete combustion of fossil fuels, and is primarily composed of particulate matter and sulfur dioxide. The other is so-called "photochemical smog," caused by the chemical interaction of sunlight and the emissions from petroleum combustion. In that reaction, nitrogen oxides and volatile organic compounds produce increased ozone in the troposphere. While the famous London smog of the mid-20th century was primarily of the first kind, and the Los Angeles smog of the late 20th century was primarily of the second kind, today's Beijing smog is a mixture of the two.

Ozone absorbs sunlight and, as such, it is a greenhouse gas although it is also much shorter lived in the atmosphere than CO_2. On the other hand, sulfur dioxide forms particles that reflect sunlight, and thus acts as a cooling agent on the atmosphere. Thus whether removing smog to improve air quality helps or hurts when it comes to climate change depends on what kind of smog you are removing. So scrubbers actually result in global warming (by removing sulfur dioxide), while catalytic convertors (that remove the precursor chemicals of ozone) have the opposite effect. Thus the relationship between actions to improve air quality and the reduction of climate warming is far from straightforward. At least it is unless you try to improve air quality by simply taking coal plants offline since then you not only remove sulfur dioxide but you also reduce carbon dioxide output. But, of course, for a government already ambivalent about trading the rate of economic growth to avoid long-term climate change, simply scrubbing the flue gasses of coal plants to improve short term air quality is going to be an obvious policy preference.

When it comes to photochemical pollution, any reduction here also helps reduce the risks of climate change, but the problem is that catalytic convertors do not completely remove the precursor chemicals for ozone and thus can't offset the effects of the rising rate of growth of automobile ownership, let alone reduce those effects. In the long run, alternative fuel vehicles may solve this problem, but in the short run these alternatives are far from ready as a cost effective alternative deployable at a scale commensurate with the demand for private vehicles.

V

Smog is not the only source of poor air quality of course, and while many sources of it contribute nothing to greenhouse gas output, the story is very different when it comes to black carbon which is the second most salient source of poor air quality other than smog. (Although it is not a major source of the air quality problems in Beijing. See Zheng et al. 2005.) It is estimated that two million people die annually caused by the effects of black carbon and related emissions due to indoor burning of fossil fuels alone (World Health Organization 2011). Recent research (Bond et al. 2013) has suggested that black carbon is responsible for a much higher degree of global warming than previously thought: 1.1 watts per square meter of radiative forcing, second only to the 1.7 watts per square meter produced by CO_2 (Bond et al. 2011). And taken together with methane, black carbon is now thought to constitute over 40% of current anthropogenic sources of radiative forcing (National Oceanic & Atmospheric Administration n.d.).

Black carbon is primarily caused by forest and savanna burning and by residential biofuel used in traditional cooking. The remaining sources are from diesel engines and industrial uses. On the other hand, the primary sources of methane are mining of fossil fuels, livestock, and landfills. Could these sources of climate warming be more easily curtailed than CO_2 without any significant tradeoff in limiting economic growth? Shindell et al. (2012) argue that implementing measures to control methane and black carbon output has the potential to limit global warming to 2.25 °C by 2070 without any measures to restrain CO_2 output. Among their suggestions (based on Shindell et al. 2012, table S1) are:

1. Extended recovery and utilization of associated gas and improved control of unintended fugitive emissions from the production of oil and natural gas.
2. Reduced gas leakage from long-distance transmission pipelines.
3. Separation and treatment of biodegradable municipal waste through recycling, composting, and anaerobic digestion as well as landfill gas collection.
4. Upgrading primary wastewater treatment to secondary/tertiary treatment with gas recovery and overflow control.
5. Control of methane emissions from livestock, mainly through farm-scale anaerobic digestion of manure from cattle and pigs.
6. Intermittent aeration of continuously flooded rice paddies.
7. Diesel particle filters for road and off-road vehicles.
8. Introduction of clean-burning biomass stoves for cooking and heating in developing countries.
9. Replacing traditional brick kilns with vertical shaft and Hoffman kilns.
10. Replacing traditional coke ovens with modern recovery ovens, including the improvement of end-of-pipe abatement measures in developing countries.
11. Elimination of high-emitting vehicles in road and off-road transport.
12. Banning on open burning of agricultural waste.
13. Substitution of clean-burning cook stoves using modern fuels for traditional biomass cook stoves in developing countries.

The great virtue of this approach is that it holds the promise of buying time to deal with CO_2 output by tackling sources of greenhouse gasses for which ready fixes are available and for which the costs of implementation are well known. Moreover, it correctly ignores land clearing by fire which is an integral part of "strategies" to promote economic growth in Latin America and Asia and will be hard to limit without providing an alternative pathway for those trying to increase their level of wealth. That said, this pathway to limiting global warming is more complex than might seem at first blush. Limiting black carbon from diesel engines and industrial uses, and methane from mining and landfills would be relatively easy to accomplish using existing technology and regulation. But attempts to promote the adoption of efficient cook stoves face cultural problems of resistance to change. They also pose a practical problem of ensuring proper adjustment of and maintenance of the equipment without which there is little reduction in pollution when compared to traditional methods. Finally, the only way to effectively limit methane from livestock at scale is to reduce livestock consumption in the face of rising demand for meat protein as household income rises in the Developing World.

These reservations notwithstanding, whatever gains can be wrung from this strategy, along with the substitution of natural gas for coal, buy us *some* time. But whether it is enough time to transition to fossil free energy depends not only on how successful efforts to constrain methane and black carbon are, but also the relative cost of fossil free fuel over fossil fuel as well as, of course, the rate of its deployment. All of these unknowns reinforce the idea that it would be foolhardy to wait to deal with carbon output. So while we might wonder whether restraining methane and black carbon would buy enough time for the debate between optimists, pessimists, and agnostics about when China's carbon output will peak to be settled by actual economic data, not acting now seems imprudent.

That said, we have canvassed seemingly powerful reasons causing both China and the United States not to act now, driven by both questions about the cost of doing so and the rate at which the future is discounted. Given their failure to act, it is surprising to find that others have acted, and acted on their own. Just why they chose to do so is puzzling because their actions seem to be inconsistent with all of the considerations we have examined so far about the prospects for unilateralism to work. Whether understanding why they did what they did can provide a model for others remains to be seen.

References

Ben-Menahem, Yemima 1997. "Historical Contingency," *Ratio*, 10: 99–107.

Beutler, Brian 2010. "Graham: I'll Filibuster My Own Climate Bill Unless Reid Drops Immigration Altogether," http://talkingpointsmemo.com/dc/graham-i-ll-filibuster-my-own-climate-bill-unless-reid-drops-immigration-altogether, accessed July 3rd, 2013.

Bond T. C., S. J. Doherty, D. W. Fahey, P. M. Forster, T. Berntsen, B. J. DeAngelo, M. G. Flanner, S. Ghan, B., D. Koch, S. Kinne, Y. Kondo, P. K. Quinn, M. C. Sarofim, M. G. Schultz, M. Schulz, C. Venkataraman, H. Zhang, S. Zhang, N. Bellouin, S. K. Guttikunda, P. K. Hopke, M. Z. Jacobson, J. W. Kaiser, Z. Klimont, U. Lohmann, J. P. Schwarz, D. Shindell, T. Storelvmo, S. G. Warren, C. S. Zender 2013. "Bounding the Role of Black Carbon in the Climate System: A Scientific Assessment," *Journal of Geophysical Research-Atmospheres*, 118 (11): 5380–5552.

Bond, T., C. Zarzycki, M. Flanner and D. Koch 2011. "Quantifying Immediate Radiative Forcing by Black Carbon and Organic Matter with the Specific Forcing Pulse," *Atmospheric Chemistry and Physics*, 11, 1505–1525.

Buckley, Christopher 2013. "China Issues Proposal to Narrow Income Gap," *New York Times*, February 5th, www.nytimes.com/2013/02/06/world/asia/china-issues-plan-to-narrow-income-gap.html?_r=0, accessed April 17th, 2013.

D'Souza, S. 1995. The Montreal Protocol and Essential Use Exemptions," *Journal of Aerosol Medicine*, 8 (Suppl 1): S13–S17.

Lizza, Ryan 2010. "As the World Burns," *New Yorker*, October 11th, www.newyorker.com/reporting/2010/10/11/101011fa_fact_lizza#ixzz2PE2d7Z1p, accessed April 1st 2013.

Morrison, Terri and Wayne Conaway 2006. *Kiss, Bow, or Shake Hands*, Avon, MA: Adams Media.

National Oceanic & Atmospheric Administration n.d. *Radiative Forcing of Climate by non-CO2 Atmospheric Gases What is Radiative Forcing of Climate by Trace Gases?*, www.esrl.noaa. gov/research/themes/forcing/, accessed November 1st, 2013.

Peng, Wuyuan 2009. "The Evolution of China's Coal Institutions," *Working Papers Series #86 of The Program on Energy and Sustainable Development, Stanford University*, http://iis-db. stanford.edu/pubs/22612/PESD_WP_86.pdf, accessed July 1st, 2013.

Shindell, Drew, Johan C. I. Kuylenstierna, Elisabetta Vignati, Rita van Dingenen, Markus Amann, Zbigniew Klimont, Susan C. Anenberg, Nicholas Muller, Greet Janssens-Maenhout, Frank Raes, Joel Schwartz, Greg Faluvegi, Luca Pozzoli, Kaarle Kupiainen, Lena Höglund-Isaksson, Lisa Emberson, David Streets, V. Ramanathan, Kevin Hicks, N. T. Kim Oanh, George Milly, Martin Williams, Volodymyr Demkine, David Fowler 2012. "Simultaneously Mitigating Near-Term Climate Change and Improving Human Health and Food Security," *Science*, 335 (Jan): 183–189.

United Nations 2011. *Report of the Conference of the Parties on its Seventeenth Session, held in Durban from 28 November to 11 December 2011*, http://unfccc.int/resource/docs/2011/cop17/eng/09a01.pdf, accessed on May 1st, 2013.

Verstappen, Stefan 2008. *Chinese Business Etiquette: The Practical Pocket Guide*, Berkeley: Stone Bridge Press, Kindle Location 636.

Victor, David 2011. *Global Warming Gridlock*, Cambridge: Cambridge University Press.

White House 2001. Text of a letter from the President to Senators Hagel, Helms, Craig and Roberts, http://georgebush-whitehouse.archives.gov/news.releases.2001/03/20010314. html, accessed July 8th, 2013.

World Health Organization 2011. *Air Quality and Health*, www.who.int/mediacentre/factsheets/fs313/en/index.html, accessed April 23rd, 2013.

Zheng, Mei, Lynn G. Salmon, James J. Schauer, Limin Zeng, C. S. Kiang, Yuanhang Zhang and Glen R. Cass 2005. "Seasonal Trends in PM2.5 Source Contributions in Beijing, China," *Atmospheric Environment*, 39: 3967–3976.

9

GOING IT ALONE

On September 27th, 2006, Arnold Schwarzenegger signed California's AB32 into law which set in motion the promulgation of state wide regulations to reduce greenhouse gas emissions to 1990 levels by the year 2020. At the time, California was the eighth largest economy in the world. Large, but not large enough to force others to follow its lead by imposing import bans on those who did not do so. And indeed, with the exception of imported electricity, the legislation reflected no aspiration to catalyze others into action except by example.

In one sense California was not alone in implementing this legislation. New Jersey enacted legislation in 2007 and the United Kingdom did so in 2008. But in a more important sense, all of these represented unilateral actions relative to their economic neighbors and major competitors. Quite apart from the considerations of the last chapter, economic logic alone would seem to militate against such action: increasing the costs of production relative to your competitors creates an economic advantage for them. That advantage should be reflected in a price signal in the market which should prompt a shift in market share. Now there are many simplifying assumptions underlying this sort of argument. But the core claim is very much at variance with an argument commonly made by proponents of the legislation in all of these cases; namely, that far from costing anything economically, climate legislation can be expected to boost economic growth and wealth for those who enact it as compared to their competitors.

Van Jones, who briefly served as Special Advisor for Green Jobs, Enterprise and Innovation at the White House Council on Environmental Quality in the Obama administration, writes that:

> If we are going to beat global warming, we are going to have to weatherize millions of buildings, install millions of solar panels, manufacture millions of

wind-turbine parts, plant and care for millions of trees, build millions of plug-in hybrid vehicles, and construct thousands of solar farms, wind farms, and wave farms. That will require thousands of contracts and millions of jobs – producing billions of dollars of economic stimulus.

(Jones 2009, Kindle Locations 292–286)

Claims like this ignore three important considerations. First, even assuming the new jobs are not located abroad, for every green energy job, we need to ask how many fossil fuel jobs we sacrifice. The answer is far from obvious and depends on a number of issues. For example, while building a green energy infrastructure is labor intensive, running it, once completed, is not as compared to the ongoing enterprise of mining coal. Second, absent a revenue neutral policy, policies to drive up the cost of carbon to stimulate a switch to green energy reduce the purchasing power of consumers. Third, these claims ignore the potential effects on the competitiveness of an economy instituting such changes unilaterally.

During the public debate leading up to the enactment of AB32 in California arguments about the economic effects of the legislation, while not dominant, nonetheless played some role because Arnold Schwarzenegger, as a Republican governor, had to answer concerns of his party's business constituency. The most serious treatment of the issue was produced by Margo Thorning (2006) on behalf of the American Council for Capital Formation, to which Michael Hanemann (2006) of the California Climate Change Center at U.C. Berkeley replied.

Thorning's assessment of the negative economic effects of AB32 was made up of three key claims: first, that it would cause job loss. Second, that an emissions trading system with mandatory emissions caps would slow growth. And third, that it would disadvantage California relative to other states and competitors. These claims rested of a mix of political analysis, economic modeling, and economic theory.

The claim that climate legislation would cause job loss was based entirely on economic modeling, with an assumption that the industries hardest hit by climate policies would be oil refining, the chemical industry, and utilities. Thorning relied on a 1998 macroeconomic aggregative "top-down" analysis conducted in 1998 that examined the effects of the adoption of the Kyoto Protocol. That analysis projected a reduction of 3% in California's gross state product by 2010, a fall in income of $1,600 for a family of four, 278,000 fewer jobs and a $14.3 billion loss of tax revenue. It should be noted that the Kyoto target was 7% below 1990 emissions by 2010, while AB32's target was 1990 levels by 2020. Still, that difference notwithstanding, Thorning assumed that the analysis (if correct) could be treated as indicative of the direction of the effects of capping carbon emissions.

The idea that climate legislation would slow growth was based on an argument about the effects of uncertainty on the rate of business investment. Proponents of climate legislation often argue that setting long-term mandatory targets for emissions gives business a predictable environment in which to engage in long-term planning with an appropriate degree of confidence. However Thorning argued that mandatory caps have the exact opposite effect because, even if they set a stable target, two unknowns create more, not less, certainty. The reason why is that reaching those targets is a function of the underlying rate of economic growth and associated energy use. Uncertainty about these, she argued, creates uncertainty about the return on investment on any new energy technologies, since they might not prove adequate as state emissions trajectories are adjusted to accommodate unanticipated growth in energy consumption.

Projections of job loss due to the effects of climate legislation came from two different sources. One is the depressive effect of higher energy costs on the economy that essentially acts as a tax and, like any tax, slows economic growth. But the other source is the projected "leakage of jobs" were California to implement climate legislation unilaterally. Such job leakage can result from actual jobs being "exported" across state lines or by jobs that would have come to California, but for the higher energy costs, that instead go elsewhere.

Now of course none of this argumentation in 2006 anticipated the recession of 2008. The debate about the economic impact of attempting to reach proposed targets of AB32 took place against a background of certain assumptions about the prospective underlying growth rate. With (almost) no growth, realizing the goals of AB32 to date has been (deceptively) easy, which is not to say it will continue to be so as growth resumes. But, be that as it may, if we bracket out retrospective knowledge, looking forward from 2006, what were the counterarguments to Thorning's analysis?

When it came to overall job and income losses, Hanemann (2006) argued that the analysis of the effects of implementing Kyoto could not be treated as a reliable indicator for the effects of implementing AB32 because of the difference in the proposed timelines, which allowed for a longer period of adjustment. But more importantly, Hanemann reported competing "bottom up" disaggregative models produced more optimistic results than those reported by Thorning because they allowed for greater elasticity of substitution.

Hanemann conceded that a cap on carbon was also a cap on growth, but only if you assumed a fixed rate of emission intensity, that is emissions per dollar of gross state product. But Hanemann argued that emissions intensity declined significantly over the previous three decades was largely driven by the adoption of energy efficiency standards that stabilized consumer energy consumption. He thought that ongoing reductions in energy intensity would, in and of themselves, drive down carbon intensity, even assuming carbon stayed part of the energy portfolio, above and beyond direct reductions of carbon intensity triggered by carbon caps. Now it is worth noticing the optimism of this view since it assumes the projectability of past gains in efficiency into the future at the same rate and cost (as well as ignoring onetime gains in energy intensity made by shifts in the economy from manufacturing to service industries).

When it comes to leakage, Hanemann conceded it too was a problem, but thought it could be contained by a carbon policy that incorporated embedded carbon in imported goods and services. This too is optimistic in the assumptions it makes about ease of administration, except for the most straightforward case of imported energy produced out of state using fossil fuels. Calculating the embedded carbon in other imported products would be a daunting task in and of itself, let alone for one state to take on, on its own. Moreover, the proposal also ignores leakage of the production of goods in California for export to other states and countries that might move out of state to jurisdictions without comparable regulations.

Still, whatever reservations may be raised about the economic analysis of the costs of unilateral action to stem greenhouse gasses both pro and con, what is noteworthy about the history of early legislation in California is the insignificant role such arguments played. The same is true in the case of United Kingdom. In each case the legislation succeeded with remarkably little opposition. To understand why, we need to examine the political rather than the economic issues that were in play in each case and how economic arguments were shaped to conform to political needs. When it comes to the political, what is striking is the fact that in California and the United Kingdom, it was conservatives who initiated the legislative process and, in each case, doing so (irrespective of their motives) helped them to position themselves toward the center of the political spectrum.

II

Fifteen months before the enactment of AB32, Schwarzenegger had signed an executive order calling for a reduction of greenhouse gasses to 1990 levels by 2020 and 80% below 1990 levels by 2050. As an executive order, rather than legislation, the action was inherently weak since it could be overridden by a subsequent executive order. Moreover, although the targets it set were ambitious by including the 2050 goal, they were just that – targets, rather than legally binding limits. Nonetheless, the executive order had the political effect of catalyzing efforts by Democrats in the legislature to extend California's existing tough auto emission standards to more far-reaching climate legislation.

The origins of Schwarzenegger decision to promote climate policies remain unclear, but, whatever else, they were not a result of any concerted political effort by the environmental constituency. Rather, he seems to have arrived at the idea through the influence of family members, especially Robert Kennedy Jr. But whatever the cause, Schwarzenegger's embrace of the issue won him favor with a majority of voters in a state that had been increasingly dominated by Democrats. In contrast, in the United Kingdom, David Cameron, who was contesting for leadership of the Conservatives, was clearly looking for ways to "de-toxify" his party's Thatcherite image. Efforts by British environmental groups to organize community support and draft model legislation on climate provided him with a ready-made issue to do just that.

If Schwarzenegger's executive order was weak as compared to the legislative action that eventually followed it, Cameron's efforts were of necessity even weaker since his party was in opposition. But they nonetheless represented a significant change for his party which had just lost an election and whose leadership he was now contesting. The Conservative Party election manifesto for the May 2005 election had voiced modest ambitions on climate policy:

> To ensure Britain plays its part in combating climate change, we will phase out the use of harmful HFCs and deliver greater incentives to make homes more energy-efficient. Through cuts in Vehicle Excise Duty and increased grants, we will significantly reduce the cost of cars with low carbon emissions. We believe that households and businesses should recycle an increasing amount of their waste.
>
> (Conservative Party 2005)

In the internal race for leadership of the Conservatives after the losing the election to Labour, Cameron called for a new climate policy that had three elements. First, an all-party commission to develop a policy framework. Second, the creation of an ongoing independent body to monitor and forecast performance. And finally, legislation with specific annual requirements for carbon reduction (Cameron 2005). Cameron went on to win leadership of his party and although it was the party in power (Labour) that introduced what became The Climate Change Act of 2008, it did so without significant opposition from the Conservatives and a great deal of behind the scenes cooperation.

California's AB32 and the UK's Climate Change Act differed from each other in one important respect. While the UK legislation set both 2020 and 2050 goals, the California legislation only did the former despite the inclusion of both in Schwarzenegger's executive order. Nonetheless, they shared much in common when it came to legislative strategy. Both bills bought some time at the front end by either delaying implementation of actual caps or by setting modest early targets to avoid causing any immediate pain to voters (and thereby incurring their wrath). At the same time, both bills assigned authority to agencies to establish trajectories for their respective economies to have a realistic chance of reaching their targets, and did so in a way that would make it politically costly for legislators to overturn. In the case of California, which has a longstanding tradition in which state agencies are held in high regard, responsibility for setting targets was assigned to the state's Air Resources Board. In the case of the UK the legislation created a new agency, The Committee on Climate Change, to plan prospective carbon budgets. In both cases these arrangements served to shift blame for unpopular policies away from politicians themselves but, at the same time, make it hard for them to water down those policies. In the case of California the only way to do so would be to repeal the legislation or defund the Air Resources Board (although the legislation authorizes the governor to adjust the deadlines for the implementation of regulations set out in the bill "in the event of extraordinary circumstances, catastrophic events, or threat of significant economic harm"). In the case of the United Kingdom, the arrangements are more complex. The Committee on Climate Change is charged with drawing up 5-year carbon budgets which must be presented for approval to Parliament. But once approved, amendment of a budget is only authorized under specified circumstances outlined in the legislation that would force the government to justify its actions.

What these elements of both bills indicate is a seriousness of purpose about the legislation designed to make its implementation resilient. After all, climate legislation is politically fraught, in a way that most legislation is not, in that it does not call for a one-off action, nor does it call for a fixed set of policies to be put into place. Instead, it mandates a regime of increasing costs and sacrifice for voters, notwithstanding that it is for the greater good. As such, it is not the kind of legislation that voters can be expected to accommodate to over time, whether or not it provokes initial outcry.

This fact, taken together with the limited value of going it alone in terms of the global reduction of carbon output and the unknown economic costs, make the decision to proceed with the legislation all the more politically puzzling. Were there other factors included in the political calculus? If you ask those involved, in both California and the United Kingdom, they give two very different sets of reasons to explain their actions, one economic and the other cultural.

The economic argument was driven by a straightforward assumption that long-term regulation of carbon output was bound to be adopted soon, if not by the world as a whole, at least in the United States and the European Union. (I say "long-term" since the European Union had a short-term plan in place governing emissions through 2020 at the time.) As such, "getting out in front" was seen as yielding economic benefits in the experience it would give in the development and organization of decarbonization. The resulting competitive advantage over those that followed was seen as outweighing any economic costs of having higher energy prices in the meantime.

The cultural reasons were more subtle. Politicians in California take pride in the fact that the state has a history of regulation that is stricter than the rest of the United States. In some cases that regulation has proved to be a model the federal government has adopted. But in others, the sheer relative size of California's economy has meant that its technical standards have been adopted de facto by other regions of the United States as a function of market rationalization of the kind discussed earlier. Thus while California has higher auto emissions standards than other parts of the country, all Volvos sold in the United States conform to California's standard because it is cheaper to produce them that way irrespective of which state they are sold in.

The United Kingdom does not bear a similar relationship of relative size to the European Union that California has to the United States. But if anything, its politicians embrace the idea that theirs is a country whose role is to set a course for others even more than politicians in California. As a member of the House of Lords put it to me, "We led the World in the abolition of the slave trade so it is only right that we should lead on limiting climate change." This embrace of the perceived duty of moral leadership dovetails with a broader trope about leadership among British politicians that dates back to the end of the Second World War: the search for a national mission that would prompt deference to, and recognition of, Britain as a world leader.

These kinds of rationalizations for political action (both economic and cultural) are the stock and trade of political discourse. But it would be naïve to treat them as just talk, as no more than rationalizations provided to justify political choices embraced out of self-conscious narrow self-interest. True, some political strategists may see things that way, especially after the fact. But to treat political leaders as making choices in such a cravenly cynical way is to fail to do justice to the dynamics of political decision making and to thereby limit one's understanding of it. Be that as it may, in the case of climate legislation, in both California and in the United Kingdom, we want to not only know how this legislation came about when it has not done so elsewhere, but also how robust or long lived this legislation is likely to be.

III

In the case of California, this is in a sense an empty question. For the decision to limit the legislation to 2020 goals alone, instead of both 2020 and 2050, and the slowdown of the recession of 2008, has made the trajectory toward realizing the goals of the legislation relatively painless to realize. What happens after that is an open question since there is no legislative authority beyond 2020, except a prohibition on relaxing the 2020 provisions themselves.

That legislators in California chose not to embrace 2050 goals in their legislation in the first place is one of the enduring mysteries about their legislation. All other things being equal, the further out in time implementation of legislation is, the less risk legislators face. By 2050 those enacting the legislation would be out of office, if not dead, and, so, well insulated from voter anger directed toward them. If they had reason to worry about the political cost of the legislation, it ought to have been the near-term consequences that should have concerned them, that is the 2020 goals. Key legislators and their aides who moved the bill give two arguments to explain their decision, neither of which are particularly satisfactory. The first account revolves around missing support for long-term legislation from center Democrats whose votes were needed to pass the legislation. These legislators primarily represented inland agricultural interests and were notably lacking in their coastal colleagues' longstanding fervor for the regulations that underwrite California's reputation as a leader in setting environmental and air quality standards. But even if this is the case it does not explain why they favored 2020 regulations over 2050. The second account is equally puzzling. It rests on the claim that while legislators saw a plausible way to get from here to there when it came to the 2020 goals, relying on existing technology, they saw no way to do so for the 2050 goals. If this account is correct, it prompts the question of why that would matter to legislators. More importantly, it ignores the catalytic role of such legislation in setting a regulative environment in which there are clear economic rewards for innovation.

Maybe there were other reasons. As one observer suggested, the California legislators may just have assumed that beyond 2020 climate regulation would be a federal concern. Or they may have assumed that, even if implementing a 2050 goal was easier than a 2020 goal, doing both was harder than a 2020 alone. But in the end, perhaps looking for a reason why this legislation was so limited is too limited in itself. Looking back on the legislation, one of the governor's key aides who oversaw the legislation expressed disbelief that it did not in fact include the 2050 goals. The 2050 goals may have just got lost in the give and take of legislative writing without anyone quite realizing it and the rest is ex post facto rationalization.

In contrast in the United Kingdom, with 2050 goals and the planning timetable dictated by the legislation, The Committee on Climate Change has already recommended four carbon budgets that cover the period up to 2027. That reaches far enough in to the future to offer a test of the United Kingdom's commitment to the enterprise as some of the real costs of it begin to bite. Of these, by far the most significant, both economically and politically, is the power generation sector which calls for a significant expansion of nuclear energy which is to be paid for by a combination increased rates levied on end users and government subsidies. That fact has combined with two other factors to significantly weaken enthusiasm for the legislation particularly in the ruling Conservative Party.

First, quite aside from the fact that, as for California, the 2008 recession made it much easier to achieve 2020 goals than anticipated because of the economic slowdown, using 1990 as the base year for climate calculations was particularly favorable for the United Kingdom. Starting in 1993, as a result of privatization and regulatory changes, the protected status of coal was weakened and the price advantage of (plentiful North Sea) natural gas could be exploited. That fact combined with the relative speed with which natural gas turbines could be built to produce a transformation in the United Kingdom's electrical energy generation portfolio in the decade of the 90s, with natural gas going from .7% to 34.5% by 2011, while coal fell from 65.3 to 34% during the same period (Department of Energy & Climate Change 2013). But with this low-hanging fruit picked, there is no painless way forward to achieve the next rounds of reductions needed to be on track to reach the 2050 goals.

Second, the increased costs associated with staying on that track has ignited a debate about the value of mandatory caps that is much more charged than the debate about it at the time of the original legislation. The seeming indifference of business interests at the time of the original legislation remains somewhat of a puzzle as much in the United Kingdom as in California. But in fact, in both cases, heavy manufacturing requiring massive amounts of energy are few and far between in their respective economies, with the exception of cement production. While utilities are of course energy intensive, they operate under regulated authority to guarantee that increased costs of generation can be passed on to consumers. Moreover, Thorning's (2006) arguments notwithstanding, in general, businesses favor a predictable legislative regime (even if it is not optimal) over alternatives because there are fewer unknowns that need to be incorporated in long-term models. Finally, we should not discount the role of state boosterism and national chauvinism in shaping the attitudes of corporate leaders as much as politicians.

That said, as the real costs of climate legislation have begun to become more apparent, as the fourth United Kingdom Carbon Plan has been rolled out, corporate concerns have become more strident, all the more so given the fact that the much anticipated harmonization (beyond 2020) between EU climate regulations and United Kingdom regulations has not taken place, thus frustrating the United Kingdom of the perceived benefits of its role as "leading."

If business stood on the sidelines during the initial legislation only to become more concerned during the current stages of implementation, the same cannot be said of the Treasury Department in the United Kingdom. In the run up to the legislation, the Treasury opposed unilateralism of emissions targets and mandatory carbon caps as well as renewable targets (although these were under consideration independent of the climate legislation itself), and it has continued to do so. While unilateralism became moot for the first three carbon budgets because the European Union adopted comparable targets in December of 2008, it is a live issue when it comes to the fourth carbon budget (2023–2027).

The Treasury's opposition to these policies largely revolved around concerns about the United Kingdom's competitiveness. But that was not the only consideration. The key to the legislation, as enacted, was the concept of multiyear binding carbon budgets which effectively institutionalized the priority of climate policy over other objectives. In this respect the Treasury had two reasons for its opposition. First the legislation limited its range of options in managing economic policy. And second, in doing so, it forced the Treasury to yield some of its power to a competing institution; namely, The Committee on Climate Change. Notwithstanding its concerns, the Treasury was overruled on all of its objections by the ruling Labour Party. However under the Conservatives now in power, the Treasury's concerns have had a more sympathetic hearing. There is of course an irony in this in that it was the Conservatives who began the push for the United Kingdom's climate policy legislation in the first place. But with the next potential leader of the party, the current Chancellor of the Exchequer, skeptical about its value, and eroding support for it on the party's backbenches, significant weakening of carbon budgets going forward is increasingly likely.

III

When Governor Jon Corzine of New Jersey signed the Global Warming Response Act on July 6, 2007 (State of New Jersey 2007), New Jersey had in place the most far-reaching greenhouse gas legislation in the world. For the state with the highest number of superfund sites in the country it was a pleasantly surprising position to be in. Corzine's staff was proud of the fact that New Jersey was not following California but ahead of it with a legally mandated goal of an 80% reduction of greenhouse gasses by 2050. The fact that no nearby state had such legislation in place seems not to have deterred the governor or the legislature.

Corzine was inaugurated as governor of New Jersey in January of 2006, but the idea of climate change legislation had played a relatively small role in his campaign, as did his environmental campaign commitments more generally. In a speech on the environment in the campaign (given on October 10, 2005), Corzine's pledged to reduce energy consumption by 20% and grow renewable energy resources by the same amount. The first part of this commitment did not specify from what base line the reduction would be measured. The second was ambiguous about whether the growth would be a 20% addition to the existing renewable resources or to 20% of all energy production. Less than a year after his inauguration though, Corzine signed an executive order with much more unambiguous and ambitious goals for the state. Executive Order #54 called for emissions targets for greenhouse gasses at 1990 levels by 2020 and 80% below 2006 levels by 2050. Less than 6 months later the same targets were signed into law giving them robust standing beyond the term of the governor (in contrast to the executive order). Moreover, the "targets" of the executive order became more prescriptive in the language of the bill which states that "No later than January 1, 2020, the level of Statewide greenhouse gas emission shall be reduced to, or below the 2020 limit" and so too for the 2050 provision.

How much did perceived economic interest drive this legislation? This much is certainly true: after the fact both the Governor and the prime mover (Assembly-woman Linda Stender) argued that the bill would promote economic growth that would counterbalance any "costs" to the economy that it might create. But as in the case of California and the United Kingdom, the transcripts of the hearings and other supporting materials reinforce the perception that neither economic cost nor benefit was talked about much before the legislation passed.

Nor was the legislation a response to voter interest that might be seen as yielding widespread political benefits to the actors involved. Indeed the record from the hearings and supporting material give scant support to the idea that there was any sort of groundswell of voter interest in the bill, let alone prior strong support for the issue. As in the other cases we have been examining, in interviews, all of the critical players involved, from the governor on down, readily suggested that they thought passing the bill was the right thing to do and something that needed to be done, even if it only set an example and precedent for other states and the nation. (Although this is not to say that these high-minded interests were not consonant with a concern not to alienate the environmental groups in the state, which were seen as an important constituency for the Democratic Party organizationally, if not in terms of numbers of votes.)

New Jersey's legislation may have been stronger than California's in its targets, but it was critically weaker in two important respects. First, although the legislation mandated binding reductions by 2020 and 2050, it imposed no trajectory from here to there. Nor did it impose any binding mechanism by which such a trajectory was to be developed and put in place. Instead, the legislation called for the state's Department of Environmental Protection (and other state agencies) to recommend measures to reduce emissions by 2008 for the 2020 limit and 2010 for the 2050 limit. Those recommendations were to go to the governor and state legislature, but the legislation was totally silent beyond that. In theory then, the governor and the state legislature could choose not to act until the last minute and only then adopt an, albeit preposterously unrealistic, trajectory (in 2019 for the 2020 limit and 2049 for the 2050 limit as it were) while still remaining within the letter of the law. In fact, to date, neither the governor nor the legislature have chosen to act, and thus the legislation remains, for all intents and purposes, essentially unimplemented.

The failure here is not just a product of a critical weakness of the legislation in comparison to California and the United Kingdom which bound themselves to a mechanism to develop and imposed a trajectory. As we have seen, legislation in California and the United Kingdom had its origins in the Republican and Conservative parties respectively, which functionally turned climate legislation into a consensus issue of the left and right, thereby removing it as an arena for interparty competition. But in the case of New Jersey, the legislation was moved by a Democratic governor and legislature. As such it remained an issue of contention between the parties and, as soon as a Republican (Chris Christie) came to power the legislation was consigned to oblivion.

IV

The recession of 2008 (and Britain's "dash to gas" in the 1990s) made realizing a plausible trajectory for 2020 goals much easier than initially conceived by promoters of climate legislation. In this sense the true test of their unilateralism is still to come. In the case of California legislative authority to set carbon budgets beyond 2020 will be a sine qua non, while in the United Kingdom resisting attempts to revise the fourth carbon budget will be a crucial test. But even if these efforts are successful, what can we learn from these efforts for the rest of the world? Do they offer a model that other countries could emulate? And do they demonstrate the possibility of bootstrapping our way to a global regime one country at time?

There are three reasons to think these efforts can't be a model for the rest of the world.

To spend time in both California and the United Kingdom is to be impressed with how little manufacturing there is in their economies today. In the case of California that may not seem surprising since we carry an image of the state as clean and pristine to begin with, even if this romantic view is bellied by the economic history of the state in which both the oil industry and the production of military hardware played a major role. When it comes to the United Kingdom however, the contrast is more straightforward. Its economic history is very much at the root of our idea of it as the center of the industrial revolution, so the idea of it as postindustrial does come as a surprise.

But if heavy industry and significant manufacturing no longer play a role in these economies, it would of course be naïve to think they don't rely on them. It is just that these activities occur elsewhere, chasing the lowest wages and cost of manufacturing around the world. The never ending movement of heavy industry as less developed economies become more developed, and even less developed economies begin to industrialize, highlights the weakness of viewing California and the United Kingdom as offering generalizable models of climate regulation. For implementing carbon caps in an advanced economy, with limited manufacturing and heavy industry, is much less costly and disruptive than it is in a less developed economy. Nor is the difference between such economies readily eliminable. Not every economy can forgo heavy industry unless the world economy is willing to forgo what these industries produce altogether. Instead, the only way to render climate regulation in California and the United Kingdom as generalizable models would be to incorporate the embedded carbon in all of their imported goods within their carbon caps and by the same token provide offsetting allowances to the exporting countries despite the previously discussed challenges of doing so. But legislating carbon caps this way would have been a much harder political burden for legislators to carry since the cost to their citizenry would have been much higher.

Still, perhaps those costs might have been offset by the perceived economic and political benefits of being first and leading "ahead of the pack." But if that were the case, these could not then be generalizable models for the rest of the world any more than all of the students of Lake Woebegone can be above average.

V

Is there a way forward then? I have argued that the prospects for an agreement between large economies like the United States and China that could force others to follow them are dim. At the same time, the idea that smaller economies like California and the United Kingdom can lead by example should not inspire confidence. Are we then doomed to sleepwalk toward potential disaster? Some may hope that nature will be kind enough to give us clear instances of catastrophic consequences of climate change soon enough for us to take collective heed and act together decisively. But so far at least, that has not happened. The increased frequency and intensity of disruptive weather has either not been clearly enough tied to climate change or not experienced as severe enough to make a difference. If we indulge in thought experiments to ask what it would take, it is tempting to assume a series of catastrophes that uniformly affected all parts of the world at the same time. But that is one thing disruptive weather is almost guaranteed not to provide us with. Yes, tornadoes and hurricanes may increase in number and frequency in the United States, monsoons may increase in intensity in India, and drought may worsen in Africa, but even in the statistically unlikely event that these were to happen simultaneously, they would leave billions of people and the majority of countries in the world unaffected and, as such, likely unmoved.

This pessimistic scenario suggests that if we act, we will only be moved to do so collectively when the accumulated effects of climate change are clearly attributable to it and widely distributed enough in time and place to affect most of the globe. What then, if anything, is to be done?

References

Cameron, David 2005. "Change Our Political System and Our Lifestyles," *The Independent*, November 1st, www.independent.co.uk/voices/commentators/david-cameron-change-our-political-system-and-our-lifestyles-513423.html, accessed May 9th, 2013.

Conservative Party 2005. *The Conservative Party Manifesto*, http://news.bbc.co.uk/2/shared/bsp/hi/pdfs/11_04_05_conservative_manifesto.pdf, accessed May 8th, 2013.

Department of Energy & Climate Change 2013. *Historical Electricity Data: 1920 to 2012*, https://www.gov.uk/government/statistical-data-sets/historical-electricity-data-1920-to-2011, accessed December 12th, 2013.

Hanemann, W. Michael 2006. *Review of Economic Arguments against AB32*, www.e2.org/ext/doc/Review%20of%20Economic%20Arguments_Hanemann.pdf;jsessionid=C9E33AE935560D22E94D756198366AD9, accessed May 3rd, 2013.

Jones, Van 2009. *The Green Collar Economy*, New York: HarperCollins.

State of New Jersey 2007. *The Global Warming Response Act*, www.njleg.state.nj.us/2006/bills/a3500/3301_r2.pdf, accessed July 1st, 2012.

Thorning, Margo 2006. *California Climate Change Policy: Is AB 32 a Cost-Effective Approach?*, Washington, DC: The American Council for Capital Formation.

10

WHAT IF IT IS TOO LATE?

I

If we will only be moved to act when the accumulated effects of climate change are clearly attributable to it, and widely distributed enough in time and place to affect most of the globe, will it be too late to do anything?

One thing is for sure, with regard to some changes, there will be no going back. While it might be possible to "bank" the DNA of species at risk of extinction for later reintroduction, that will be comparatively easy as compared to dealing with something much simpler than life forms; namely disappearing ice. For to reconstitute the Greenland ice sheet, or glaciers once melted, would require not just going back to the way things were, but going back to a much colder state of affairs that would be, if not as challenging for us as the hotter state of affairs to which we are heading, certainly incompatible with sustaining human life without a significant reduction in the population.

But that is not to say we could not do anything, assuming we could remove CO_2 from the atmosphere, and there is no doubt we not only could do that, but could do so using existing technology. Indeed, if we wanted to, today we could wind back the carbon in the atmosphere to pre-industrial levels, even if doing so would not recreate the world as it was then. Granted it would not prevent the knock-on effects of the changes we have already set in motion, especially when it comes to ocean ecology. But what it would do is stem new effects of ongoing increasing temperature and also blunt the effects of a sustained environment at the elevated temperatures that are already "baked in" at current levels of CO_2 in the atmosphere.

"Do what?" you might ask "And at what cost and at what risk?" Let us briefly examine the techniques that at least hold out the prospect of making a significant difference. Such interventions fall into two classes: ocean based and land based. Ocean uptake of CO_2 has the prospect of being increased in one of two ways. One is by "liming" the ocean, thereby changing its acidity which increases the rate at which the ocean can directly absorb CO_2. The other is to "fertilize" the ocean with iron, thereby increasing the prevalence of phytoplankton which consume CO_2. Land-based techniques rely on either global reforestation or artificial air filtration and sequestration of the captured CO_2.

Any intervention will have to be massive to significantly reduce the atmospheric concentration of CO_2. But, paradoxically, for land-based and air-filtration interventions, in doing so, it will also make things worse in the sense that removing CO_2 from the atmosphere will lead to a release of CO_2 from both the ocean and terrestrial sources. The reason why is that changes in the concentration of CO_2 in any one of these mediums (land, water or air) disrupts the equilibrium between it and the concentration of it in the other mediums. The balance between them has had a modulating effect on the relationship between our emissions of CO_2 to date and the increase of it in the atmosphere since our actions have caused an increase in ocean and terrestrial uptake in response. But, by the same process, the piper needs to be paid if we act to reduce the concentration. Just as ocean and terrestrial uptake absorbs (at most) half of what we add to the atmosphere in any one year, we should assume the opposite will happen in any air capture program. That is to say, up to 2 ppm of CO_2 will have to be removed for every 1 ppm of net reduction we aim to achieve.

If we look at these alternatives at the most general level, bracketing both the questions of effectiveness and potential side effects for the moment, you might think we need to worry about two factors. One is a matter of economic costs. Since the argument here is that carbon usage is a necessary condition to drive economic growth, which is needed to limit population growth, a solution that consumes too much of economic output in a non-productive way will be self-defeating. Another is a matter of energy costs. If the energy requirements to fuel a solution are greater than the energy produced by the carbon output it is capturing, then here too the approach will be self-defeating.

But neither of these really poses unavoidable obstacles. When it comes to economic costs, the high expenditures to build infrastructure and run a carbon capture program at scale would no doubt force a change in economic priorities, but that does not necessarily imply any impact on the level of economic output per se. As we saw in the earlier discussion of the net employment effects of switching from coal to solar, the answer is far from obvious and depends on a number of issues. Here the critical issues will be not just how labor intensive building the infrastructure for carbon capture would be, but how labor intensive it would be to run on an ongoing basis. The other issue is whether or not the cost of building the infrastructure was financed on a revenue neutral basis (let alone by long term deficit financial instruments) designed to avoid reducing the purchasing power of consumers. When it comes to energy costs, even if the cost of carbon capture is greater than the energy production associated with that level of carbon release, such a program would not necessarily be irrational as long as non-fossil fuels were used to drive the capture process. But then why not use that non-fossil fuel instead of the carbon in the first place? Of course if non-fossil fuel was available in the amount needed, and on the timetable needed, there would be no reason not to do so, but here we have assumed otherwise. But moreover, if we allow ourselves to be optimistic about the availability and cost of non-fossil fuel in the future, the relative cost may not cut against the seeming lunacy of using more energy in the future to clean up the effects of the use of energy now.

If economic costs and energy costs are not the crucial factors when it comes to carbon capture, three other factors are, and they stand out as potential killer objections to any program irrespective of need. One is a matter of realizability at scale. While reforestation may seem the most straightforward and "natural" way to capture carbon, the sheer amount of land mass that would be needed is far in excess of what is available even ignoring population density, agricultural land needs, and the availability of water. (For example, were an area the size of all of the United States, including Alaska, covered in trees de novo, the net annual carbon captured would be equivalent to a reduction of 3 ppm per year of CO_2.) Another is the predictability of the method. Thus ocean fertilization, even if it were possible to do at scale for the desired effect, depends on organisms to realize those effects. However those organisms are subject to evolutionary change and there is no saying how they may evolve in the new environment created by the addition of iron fertilization. (In laboratory work Sinead Collins and her colleagues (Lohbeck et al. 2013), have tracked 1,000 generations of algae in a CO_2 rich environment and observed surprising declines in uptake rates.) Finally, even if scale and predictability concerns are satisfied, a third area of concern is stability. CO_2 capture is of limited value if it leaks back into the atmosphere, and even worse, were it to do so in large amounts over a short time it could produce its own environmental shock to the ability of both individual organisms and whole ecosystems to adapt. The potential for slow leakage is clearly a concern with most ocean capture proposals. While underground storage of CO_2 certainly also poses a risk for slow leakage, concerns have also been raised about the danger it poses for catastrophic local discharges like that which occurred at Lake Nyos in 1986 killing 1,700 people (see Fleming forthcoming). However, even were such discharges to occur, to pose a serious threat they would have to be in globally large amounts. But that seems unlikely given the likely dispersal of CO_2 storage in varied locations.

That said, slow leakage of sequestered CO_2 underground poses enough of a problem to be taken seriously given the vast scale of carbon capture that might be involved. The only way to avoid this problem is to meld carbon capture to a system of mineralization that binds the carbon in stable compounds. Rau et al. (2013) have recently proposed such an approach, using electrolysis, which they argue is much more efficient than mechanical systems both qua economic costs and energy costs. Whether that turns out to be the case, along with questions of scalability, remains to be seen.

Be that as it may, the fact of the matter is that, just as we could cover the Sahara desert with solar panels to fuel our energy needs, we could do likewise for carbon capture if the perceived need made the cost acceptable, even with existing technology. But the perceived need would have to be very high given the cost. We can get a sense of this, at least on the most pessimistic view, from the recent American Physical Society study (Socolow et al. 2011) which projects the cost of capture at no less than $600 a ton of carbon dioxide. (This figure excludes the cost of transportation and sequestration, but assumes a 7% return on investment, the opportunity cost, and assumes 70% efficiency if fossil fuels are used to run the system.) Assume a system (using current technology) that absorbs 20 tons of CO_2 per square meter per annum and hence 400 tons over the projected 20 year life of a system. To absorb one megaton of CO_2 per annum you need 5 arrays each 10 meters high and 1 kilometer long each separated by 250 meters. So each 1.5 square kilometer area will absorb 20 megatons over its lifetime. Given 1 part per million of $CO_2 =$ 7.81 gigatons of CO_2, you would therefore need 390 of such 5 array units covering 585 square kilometers. Thus to reduce the atmospheric concentration of CO_2 by (say) 50 ppm every 20 years you would need to set aside and cover 29,250 square kilometers. (For comparison, Massachusetts is 27,336 square kilometers.) When it comes to economic costs, at $600 a ton, the cost of removal is $4.6 trillion per ppm of CO_2. But it is double that figure when we take into account the earlier point that reducing atmospheric concentrations of CO_2 will disrupt the equilibrium between it and the concentration of it in other mediums so that up to 2 ppm of CO_2 will have to be removed for every 1 ppm of net reduction we aim to achieve. On this basis, removing 50 ppm every 20 years thus costs an astonishing $460 trillion. It is worth keeping in mind that, assuming 3% annual real growth, cumulative global GDP will grow to $2,036 trillion by 2034. But that still means the cost of remediation is 22% of GDP. Of course, whether or not air capture is worth this cost is a function of the price of doing nothing. Moreover it is worth reminding ourselves that the cost to the United States of prosecuting World War II amounted to 42% of its GDP. Still, if spending 22% of GDP on air capture seems implausible, by scaling up the timetable, we can scale back the bite of GDP to more manageable proportions. But that does not mean costs will be insignificant if the timetable is meant to be responsive to the threats posed by elevated levels of atmospheric CO_2.

A potential alternative to direct air capture may be to achieve the same end by modifying the rate at which oceans degas CO_2 by adding base chemicals to bind the CO_2. This kills three birds with one stone by reducing the net atmospheric concentration of CO_2, and obviating the need for a separate step of sequestration while at the same time reducing the acidity of seawater. Nonetheless the scale here too is vast but maybe considerably less vast than air capture on land. (I am grateful to Gregory Rau here.)

Even if these challenges could be rendered economically plausible, there is nonetheless an undeniable sense of chutzpah that is hard to disentangle from talk about geoengineering at such a grand scale. Gardiner raises the argument that it is more of the same reckless attitude that got us into the problem in the first place when he writes that:"we could clean it (our planet) up . . . but so intent are we on continuing our messy habits, that we will pursue any means to avoid that, even those that imposed huge risks on others and involve further alienation from nature" (Gardiner

2009, 304). While Jamieson writes that even if geoengineering were successful "it would still have a bad effect of reinforcing human arrogance and the view that the proper relationship to nature is one of domination" (Jamieson 1996, 332). And yet even if we embrace such considerations, do they help us resolve the dilemma that we face? Here we are at a fork in the road. Should we turn left or right? Which will do less harm? Maybe we should not have gone down the road that we have traveled as a species. But pointing that out is not of much help. Still, Jamieson and Gardiner would likely argue that faced with such choices, there is often a technological fix. But in the long run, they counsel that making choices relying on such technological fixes will not solve the underlying problem, arguing instead that we have to find a place consistent with the limits of nature. Yet, even if we were to swear a solemn collective oath to do that, and had the collective means to follow through on such an oath, we are still where we are. We still may need to avail ourselves of technical fixes in the transition for such a transition to even be possible.

II

A program to reduce CO_2 in the atmosphere at scale would be an enormous enterprise to put in place and would have to run for many years to produce tangible results. Is geoengineering by reducing the sunlight reaching (solar radiation management) the earth a feasible, sensible, or practical stopgap measure that might be used in the meantime to limit the effects of elevated CO_2 in the atmosphere?

We do not know at this point and so it seems hardly contentious to claim that we should find out, especially since our level of ignorance reaches the most elementary aspects of the technical basics of such an intervention. For example we have theories about the concentration of sulfate aerosols that would have to be introduced into the atmosphere to reduce the temperature gains caused by CO_2, but we lack reliable experimental knowledge about the effective insertion of aerosol precursors, aerosol formation, behavior or lifespan in the stratosphere. But even if we had such knowledge to support the feasibility of such a technique, should we use it?

One objection against doing so is the risk of moral hazard (as it is, by extension, for air capture as well). But it is hard to discern why moral hazard should function as a deterrent to action here anymore than it does elsewhere. Governments regularly sponsor research and programs to counter the effects of ill-advised actions, even their own actions, that have the unintended (but ineliminable) effect of encouraging those very actions. An amnesty program for illegal immigrants may be designed to offset the effects of poor border enforcement, but in doing so it encourages more not less challenges to the border. But that is no reason not to have an amnesty program, unless the *overall* result is to make things worse than they were. All such programs involve a balancing act of benefits versus costs to arrive at the net benefit. If moral hazard arises for solar radiation management or air capture, it only matters if you think that implementation would undermine other actions to such a degree that it would lead to more *total* greenhouse gas output. And that seems far-fetched.

Be that as it may, at the moment we don't even have the knowledge to implement such a program. There is however a different worry when it comes to research about it quite aside from moral hazard, the matter of inertia. It is the kind of worry that comes to the fore when you look at the history of research programs like the U.S. missile defense shield that beget their own rationale for continued funding because of the funds spent already. But if this is a general feature (albeit a dysfunctional one) of large scale scientific research programs, unless you think no such research should take place because of it, you need to produce an argument that solar radiation management or air capture have some special exceptional features that make this concern relevant here where it is not elsewhere.

But even if these arguments do not hold sway, there is a much deeper methodological argument against solar radiation management but not air capture.

On the standard, if idealized model of science, the road to full deployment has a number of way stations, each of which offer an opportunity to assess benefit under increasingly realistic conditions. There is of course a tradeoff here. The more restrictive the realism, the less the risk, but also the less our ability to assess the benefits. Wherever this process starts, be it in the lab or in a computer model, full-scale deployment does not take place before testing in more limited circumstances, in scale, strength, or range. So in medicine, even after animal testing, we restrict the number of subjects exposed and increase the strength of the exposure in a series of steps. In medicine we can follow this procedure because of something we take for granted: our object of interest is reasonably modular or encapsulated. That is what makes it possible for us to extrapolate from experimental subjects to the population as a whole with confidence. The experimental subject is a full scale representation of the objects of interest. Of course not all of our scientific or technical interests are necessary or sometimes allow for full-scale representation at the experimental level, but in such cases, we have to at least be able to extrapolate to scale with confidence. (Think only of how airplanes are built based on wind tunnel studies of smaller scale models.) But what if the object of your interest is not modular or encapsulated? What do you do then? For that, after all, is the feature of solar radiation management. It calls for intervention on a system that lacks just this characteristic. You cannot encapsulate part of the atmosphere and it is too complex to be able to build a realistic nonvirtual model at a smaller scale. As such, it is reasonable to ask whether we could ever have a sound basis for moving to full deployment of any such proposed intervention. And if not, then why bother to even research such proposals in the first place? Before examining this question, we should pause to note that, at least in some respects, the quandary posed here is not unique to solar radiation management. While it is true that most of science and technology does deal with modular or encapsulated systems, some of our interests force us to deal with what might be termed "bioengineering," deliberate attempts to intervene to change the Earth's biology. After all, eliminating smallpox or polio is not possible without operating on a global scale.

But there is a difference here. In these cases we can study the effects of eliminating them at a local level even if to reach our goal we have to operate at a global level. Some people would argue that genetically modified crops represent a global intervention despite assurances that these can be treated as local interventions. But here too there is a difference, as we noted in Chapter Two, we have a wealth of experimental field studies to fix the probability of such interventions "going global."

There are many things about solar radiation management that we could learn from experimentation. But they are restricted to the practicality of such interventions as discussed earlier for the case of sulfur insertion. But could we ever have a basis for proceeding to deployment at scale with confidence? In a nonencapsulated system in which no scale model can be built, the only experimental model is to extrapolate from low dosage to dosage at scale. But extrapolate using what mathematical model? At least in the case of sulfur insertion, I think there are two bases for such an answer. One is theoretical, the other empirical. The theoretical one depends on arguing a case based on our general knowledge of atmospheric chemistry. The empirical one depends on our knowledge of volcanic eruptions that have produced a record of short duration high density sulfur output. But the case remains to be convincingly made. Suppose such answers are forthcoming, and suppose they support the conclusion that, even if we cannot assume linearity, there is no reason to think there is a significant risk of discontinuities, let alone runaway states. Could we ever have the confidence to move ahead by slowly increasing the insertion dosage?

Suppose we could, then the issue would be how long it would take to get reliable data. As Robock et al. (2010) have argued, proceeding with caution means collecting enough data to distinguish noise from signal. Disruptive weather occurs naturally and so collecting statistically significant data to attribute the cause of it to sulfur insertion would require a significant amount of time, and all the more so the lower the insertion dosage. The upshot of this is that a quick acting program of solar radiation management in response to a perceived climate emergency is unrealistic. The time necessary to build a large enough data base to ensure safety would not be significantly shorter than the time needed to build the infrastructure for carbon capture.

Now one might object to this last claim as follows: even if the lead time to collect data to establish safety for solar radiation management means that no quick emergency response would really be possible, if and when implemented at scale, such a program would generate a much quicker response than air capture. Air capture not only requires the construction of an enormous infrastructure, but once in place, that infrastructure will have to operate over many years to produce a significant reduction in the concentration of CO_2 in the atmosphere. On the other hand, once solar radiation management is implemented at scale, the time to reach the desired concentration is a matter of years.

That may seem to offer a chance to buy time while air capture slowly does its work, even if the waiting period to safely implement solar radiation management were to prohibit a fast "emergency" response. But there are several elements of this argument that are worth noting.

First, it is commonly accepted that there are limits on how much sulfur dioxide could be inserted into the stratosphere and thereby there is a limit to the maximum cooling effect that could be achieved. This is not just because the particles tend to grow and fall out more quickly with greater rates of insertion, but also because of the amount of acid rain produced. So benefits of such a program to buy time are limited. (For example, Alan Robock's modeling, Robock et al. 2008, finds continuous insertion of 5 million tons of sulfur dioxide per annum would produce a reduction of 0.5 to 1 °C.)

Second, given these limited benefits, and the fact that solar radiation management does nothing but buy time, we might wonder what "emergency" would create a consensus for such a program to be launched. (All the more so in the case that some have tried to make the claim that in a "genuine" emergency, the risks of immediate full scale deployment would be worth taking without testing for safety because of the global benefits the cooling would bring.) But notice that, however we define it, no such emergency is likely to be needed for an air capture program to be launched. The reason why is that air capture offers limited risk while yielding permanent benefits through the reduction of atmospheric CO_2. For whatever the potential risks of ambient carbon capture may be, the salient characteristic it has that differentiates it from solar radiation management is that it is supremely modular. You can assess the risks of both carbon capture and carbon sequestration at whatever level of granularity you desire and scale up from there based on the results you get. As such, the triggering events for consensus to occur around the need for air capture are likely to be much weaker than whatever emergency might do the same for solar radiation management. But if that is the case, then the idea that solar radiation management might be started simultaneously with air capture is unlikely to be the case, and as such, the short-term advantages it would offer are illusory.

III

Climate change may be bad for all of us in the long run, but at least in the short run, at modest levels, it produces winners as well as losers. Elevated temperatures make life easier and increase agricultural productivity the further you are from the equator, even if not so uniformly. These "winners" under climate change will be "losers" under any geoenginering scheme. It may seem self-evident that their "loss" should carry no weight but Posner and Sunstein (2008, 1569–1570) defend a principle that those who suffer losses from mitigation should be compensated by those who gain:

> Let us assume, most starkly, that the United States would lose, on net, from a climate change agreement that is optimal from the standpoint of the world taken as a whole. As a matter of actual practice, the standard resolution of the problem is clear: The world should enter into the optimal agreement, and the United States should be given side-payments in return for its participation. The reason for this approach is straightforward. On conventional assumptions, the optimal agreement should be assessed by reference to the overall benefits and costs of the relevant commitments for the world. To the extent that the United States is a net loser, the world should act so as to induce it to participate

in an agreement that would promote the welfare of the world's citizens, taken as a whole. With side-payments to the United States, of the kind that have elsewhere induced reluctant nations to join environmental treaties, an international agreement could be designed so as to make all nations better off and no nation worse off. Call this a form of international Paretianism. Who could oppose an agreement based on international Paretianism?

Perhaps. But notice that this assumes the notion of "loss" is a straightforward matter.

Assume that as the world gets hotter and Canadian farmers' crop yields increase by $x. When geoengineering is used to cool the planet, the Canadians' crop yields decrease by $x as temperature is reduced. Should their harms be given moral standing, all other things being equal? Let us assume that any historic causal responsibility of the Developed World for creating the problem in the first place plays no role. (If you like, let "Canada" name a hypothetical country in Africa that abuts Uganda but has an idiosyncratic local climate for Africa that is identical to the country we know as Canada, that is the one north of the United States.) Let us also assume that uneven benefits of efforts to cool the planet (because of variations in local conditions) play no role either. The Canadians suffer a loss as a consequence of the implementation of a policy not just less benefits than others. (In contrast to some who might claim relative harm because they benefited from cooling less than others.)

Assume that judged relative to the post-climate change environment, the Canadians suffer a loss, although judged relative to the pre climate change environment they don't. But which preclimate change environment? That of 1775? Or some point later than that? But then which? It turns out not to matter if we make some simplifying assumptions. To simplify things, assume broadly uniformly rising temperature over time (National Oceanic & Atmospheric Administration 2010) that has been uniformly good for the Canadians since anthropogenic climate change began, so that agricultural productivity has gone up for them (all other things being equal). Then, relative to any arbitrarily chosen earlier point in time and temperature, geoengineering, designed to cool the planet, will leave a Canadian farmer no worse off than he or she would have been in an environment without climate change since, at all earlier times, productivity was lower.

"So what," the Canadian objects, "the fact is that you have caused me harm and as such I deserve compensation from you. The fact that you can point to a baseline relative to which I am not harmed is irrelevant to the harm for which you are responsible." And indeed this is surely the underlying rationale that a legal approach would take to the matter. Broadly speaking, the law of torts does not place qualifications on what counts as harms, except that they be losses caused by the actions (or failures to act) of another. (For an interesting discussion of tort law as an avenue for international claims for compensation for climate change see Faure and Nollkaemper 2007.)

Do moral considerations pull in a different direction? Let's focus on the nature of the Canadians' gains instead of their harms.

Inheritance or estate taxes have a history based on the search for revenue raising opportunities. The first such tax in the United States (in 1797) was enacted to pay for a naval buildup. Other such taxes were enacted to pay for the Civil War (in 1862) and the Spanish-American War (in 1898). Practical considerations notwithstanding, arguments have also been made to justify taxes on moral grounds. Bill Gates and Chuck Collins (2009) express a common sentiment when they defend such taxes on the grounds that "some of us have disproportionately benefited from the fertile economic soil we have cultivated together." But this is an argument for income taxation as much as estate taxation. What differentiates these taxes is the idea that income is (at least in theory) earned while inherited money is not. The basis for a claim of a moral difference asserted here rests on two related notions. One is that the benefits of income are deserved unlike the benefits of inheritance. The other is that inheritance violates a principle of equal opportunity. These two ideas come together in the principle of luck egalitarianism which came up in the earlier discussion of the well. Namely, if it is unfair for some to be worse off than others through no fault of their own among equally deserving people, it follows that it is also unfair for some to be better off than others among equally deserving people. But then those who are better off under such circumstances can have no complaint if they lose their gain. On this conception, the Canadians' gain is a windfall or lucky gain and, as such, it is not a gain they have a right to. So the loss of that gain does not give rise to a legitimate claim to have been harmed.

But the problem with this conception is that it is hard to limit its reach. Luck produces all sorts of gains for people that we want to treat as legitimately "theirs" irrespective of desert. At the most extreme level, a nation state conducts a lottery to assign mineral rights for lots of land. You get lucky, I don't. To say you have no right to your gain would undermine the point of the lottery, which is to assign those rights. Yet what makes that assignment fair is not that you deserved it and I did not, but that we all had an equal chance to end up lucky. But our attitude toward the lottery of life is more nuanced when it comes to the role of luck, at least when it comes to treating bad luck differently from good luck. Good fortune makes for all sorts of inequality: some are born smart and some not. Some stay healthy and some not. Accidents befall some but not others. Yet, echoing the earlier discussion of insurance rates, except on a minimalist conception of the state, we look to social policy to level some of these playing fields, through our education, health, and other welfare policies. These help the unlucky, but they sometimes also hurt the lucky in the process, if only by depriving the luck of a comparative advantage they have over others. But that is not a "loss" we think merits compensation, and not just because compensation would be self-defeating given our social aims.

Consider an extreme case of luck. Imagine a vaccination policy that we know will damage a few while it helps most by giving them immunity against disease X. It may seem self-evident that the few that are harmed as a result of our action are those for whom we should recognize a need for compensation, even when the state causes the harm and may be legally insulated from claims against it. But suppose those damaged fall into two groups. One is made sick by the vaccine. The other is made up of those who have natural immunity to X that is (mildly) superior to that granted by the vaccine, but they lose that natural immunity when vaccinated. Now they are just like the rest of us, having the lower level of immunity that the vaccine gives. They too have been harmed, we caused that harm, but it seems unlikely that we would recognize a need to compensate them. The reason why is that we don't treat such losses based on considerations of rights. Instead, where we do compensate, it is considerations of welfare that dictate policy. And losing a lucky advantage does not count for much in such considerations. (Moreover, that it doesn't reflects that welfare is not just a function of maximizing total outcome but is sensitive to its distribution.)

On this view then, the Canadians do not deserve compensation. But notice that the reasoning is based on considerations of welfare not justice. But on those grounds, does it matter how the Canadians came to have the gain they have now lost? Suppose geoengineering is deployed in the face of a climate emergency, an emergency that requires immediate action, notwithstanding the fact that, given the planetary-wide scale of the problem, both the onset of the emergency and the geoengineering response may unfold over an extended time. State action in response to the emergency causes harm for which the state may cite the defense of public necessity. Such a defense is rooted in the notion of a welfarist idea of the good of the many outweighing the needs of a few. Still, that does not dictate that any voluntary compensation program needs to be evenhanded relative to the harm caused by the state any more than it does in the vaccination case.

I have been arguing that the Canadians are harmed, but theirs is not a harm that should form the basis for a claim that deserves recognition. But is it even a harm? Suppose my tree blocks your view of the ocean. I decide to cut it down (maybe it is diseased) and you gain an ocean view. Now you try to stop me planting a new tree, claiming it would deprive you of your gain. Bracket the property dimensions of this example and focus on harm. Once you have an unobstructed view, my planting a new tree is going to cause you harm (if only because of the lost opportunity to see the ocean). But he who taketh also giveth, I am the one who caused the gain that you now are at risk of losing. On a small time scale I have hurt you, but on a longer time scale my actions balance out and you have suffered no harm. Likewise, suppose our properties abut each other. You rebuild a retaining wall which supports my retaining wall. When you subsequently redesign your wall, mine is weakened. The redesign causes me harm. But in the larger scale of time, that harm is balanced by my previous gain. Your actions offset each other. Even without appealing to the notion of detrimental reliance, time is a crucial factor here in delimiting the number of acts. When the act of giving and the act of taking are conjoined, the distinction between the two gets lost, like a shop clerk who gives me too much change and immediately acts to correct the error.

Can we render these kinds of cases as analogical to the case of geoengineering and apply the same sort of reasoning? Both agency and time would seem to stand in the way. Suppose geoengineering were to be implemented and the Canadians complained of harm due to cooling. Suppose the world community responded thus: "you have no cause for complaint, for we warmed the planet from which you gained and now we are cooling it. You are no worse off from our actions in toto." But who is this "we"? Global warming was not caused by collective action. It was caused by the uncoordinated actions of millions of individual actors. There is here no strict identity between the actor that caused the gain and the actor who caused the loss, except in the metaphorical sense. But a metaphorical sense may not be implausible to make use of, for it echoes what happens when governments act internationally to right the wrongs of their citizens through actions like war reparations payments (see, for example, Jokic and Ellis 2001).

But what about time? The longer the Canadians adjust to, and come to rely on, a new, warmer regime, the more plausible their claim becomes. You have little complaint if you luxuriate in your view of the ocean for just 1 week between the removal of the old tree and the installation of the new one. But if it has been years, and you bought the property with the view reflected in the price, things begin to look different.

There is however a practical solution for our treatment of both the homeowner and the Canadians that carries moral if not legal force when it comes to time: to forewarn them of what lies ahead. To make clear that they should not rely on the status quo, and should take that into account in their plans.

References

Faure, Michael and Andre Nollkaemper 2007. "International Liability as an Instrument to Prevent and Compensate for Climate Change," *Stanford Journal of International Law*, 26A, http://ssrn.com/abstract=1086281, accessed October 4th, 2012.

Fleming, James Forthcoming. "Carbon 'die' Oxide: The Personal and the Planetary," in James Fleming and Ann Johnson (eds.), *Toxic Airs: Chemicals and Environmental Histories of the Atmosphere*, Pittsburgh: Pittsburgh University Press.

Gardiner Stephen 2009. "Is 'Arming the Future' with Geoengineering Really the Lesser Evil? Some Doubts about the Ethics of Intentionally Manipulating the Climate System," in S. Gardiner, S. Caney, D. Jamieson and H. Shue (eds.), *Climate Ethics*, Oxford: Oxford University Press, 84–314.

Gates, Bill and Chuck Collins 2009. "Estate Tax Breathes Life into Economy," *Oneida Daily Dispatch*, December 1, www.oneidadispatch.com/articles/2009/12/01/opinion/doc4b15e4e9a4ef7833585135.txt, accessed July 1st, 2012.

Jamieson, Dale 1996. "Ethics and Intentional Climate Change," *Climatic Change*, 33: 323–336.

Jokic, Alexander and Anthony Ellis (eds.) 2001. *War Crimes and Collective Wrongdoing*, London: Blackwell.

Lohbeck K. T., Riebesell, U., Collins, S. and Reusch, T. B. H. 2013. "Functional Genetic Divergence in High CO2 Adapted *Emiliania huxleyi* Populations," *Evolution*, 67 (7): 1892–1900.

National Oceanic & Atmospheric Administration 2010. *The Instrumental Record of Past Global Temperatures*, www.ncdc.noaa.gov/paleo/globalwarming/instrumental.html, accessed July 5th, 2013.

Posner, Eric and Cass Sunstein, 2008. "Climate Change Justice," *The Georgetown Law Journal*, 96: 1565–1612.

Rau, Gregory, Susan Carroll, William Bourcier, Michael Singleton, Megan Smith and Rogers Aines 2013. "Direct Electrolytic Dissolution of Silicate Materials for Air CO_2 Mitigation and Carbon-Negative H_2 Production," *PNAS*, 110: 10095–10100.

Robock, Alan, Luke Oman and Georgiy Stenchikov, D. 2008. "Regional Climate Responses to Geoengineering with Tropical and Arctic SO2 Injections," *Journal of Geophysical Research*, 113, D16101, doi:10.1029/2008JD010050.

Robock, Alan, Martin Bunzl, Ben Kravitz, Georgiy Stenchikov 2010. "A Test for Geoengineering?," *Science*, 237: 530–531.

Socolow, Robert, Michael Desmond, Roger Aines, Jason Blackstock, Olav Bolland, Tina Kaarsberg, Nathan Lewis, Marco Mazzotti, Allen Pfeffer, Karma Sawyer, Jeffrey Siirola, Berend Smit and Jennifer Wilcox 2011. *Direct Air Capture of CO2 with Chemicals: A Technology Assessment for the APS Panel on Public Affairs*, College.

11

FUSION

I

A few years ago I received a phone call asking me to give money to buy the freedom of a slave (in the Sudan). The cost, and sacrifice to me, would be extremely modest – no more than $50, or dinner for two. It seemed an incredibly compelling ask. Nonetheless I was reluctant to give, fearing that I would just be reducing the supply of slaves and thereby driving up the "price." That seemed to me to create the potential to draw more people into the slave trade and thereby increase the number of slaves. My fear was that freeing a slave might result in more than one person being newly enslaved and so make things worse off than they were. With the help of some economists, I found out my intuitions were misplaced. Redeeming the freedom of one slave (under most circumstances) does no harm and likely does at least some good. It turns out if I buy the freedom of 10 slaves, no more than 10 people will be newly enslaved and likely less than 10 will be enslaved. So, all other things being equal, net-net, I do more good than harm by buying their freedom (for details, see Karlan and Krueger 2007). There is a complication in this analysis. It assumes that the costs of slavery can be shared so that two people enslaved for 6 months each is no worse than one person enslaved for 1 year. But that may not be the case. Suppose systematic rape is an accompaniment of the act of enslavement. But suppose we can be sure this complication does not hold. Then ought I not to give? "No" Peter Singer might say, "you can bring about a greater good by giving your money to feed starving people." (In fact that is just what Singer did say to me about this issue in conversation.)

My goal is to examine the force of this prescription in what follows by applying Singer's challenge when it comes to the question of avoiding climate change. But in doing so, I first want to examine the problem through the lens of slavery and then extend it to the problem of climate. In this regard, I am going to be especially interested in focusing on the force of the phrase "might be able to" and in contrasting it to what we are likely to do. The core question I am interested in is this: at least when it comes to doing good, like the act of freeing a slave, even if there are alternatives that would produce greater good, what kind of factors might undermine the likelihood of those alternatives being realized? That is a quick and dirty version of the question that I am interested in. The more proper version is comparative: at least when it comes to doing good, like the act of freeing a slave, even if there are alternatives that would produce greater good, what kind of factors might undermine the comparative likelihood of those alternatives being realized over the likelihood of freeing a slave? One more twist, for we ought to not just worry that the best alternative is less likely to be realized than the second best, but that that comparative likelihood of it happening may be relatively hard to change. At least when it comes to doing good, like the act of freeing a slave, even if there are alternatives that would produce greater good, how robust are the factors that might diminish the comparative likelihood of those alternatives being realized over the likelihood of freeing a slave? So by extension for climate change the question is then this: even if avoiding climate change would produce a greater good as compared to other alternatives, how robust are the factors that might diminish the comparative likelihood of our realizing it over realizing those alternatives?

My motivation for these concerns is derived from purely anecdotal considerations: reports about the enthusiasm and interest in campaigns to free slaves show that I am not alone in having been drawn to the idea. Suppose that Singer is right and that more good would come about however if people fed the starving rather than freed slaves. But suppose further that it turns out people are more likely to be moved to give (or give more) to help free a slave than feed a starving person (notwithstanding the greater benefit of doing the latter over the former).

You might wonder why these prescriptive worries arise in the first place. Even if buying the freedom of a slave is morally permissible, is it morally obligatory? If not, isn't it just supererogatory, and if so, aren't I free to exercise my supererogatory inclinations at will? Not necessarily. Even if buying the freedom of a slave is just supererogatory act, it does not follow that I am under no moral constraints. Consider a big shot who sets up a scholarship fund for whites only. Isn't he free to give his money as he wishes? Not necessarily. Doing so may produce such profound resentment among those shut out, especially given the reasons why they are shut out, that his gift does more harm than good. But here we assumed that freeing the slave does no more harm than good. So this worry is ruled out. Still, the issue here is slightly different. Here the question is, were I to consider freeing a slave (as a supererogatory act) and in so doing, doing good, when I could do greater good by engaging in another supererogatory act (like feeding the starving) instead, may I freely choose between them? Even if competing acts are supererogatory, the choice between them may not be morally unconstrained. Indeed this is a feature of consequentialism, for consequentialism dictates how we should choose between competing claims irrespective of the status that those claims have on us. (At least it does if they have the same status. When one is obligatory and the other is supererogatory, things get more complicated, requiring an analysis of why obligatory acts trump supererogatory acts.)

II

Suppose Peter Singer sets out to improve the world by prescribing courses of action that will produce the greatest good to those that wish to listen. Singer himself, and with him, many others, would say that helping is not a supererogatory act but rather obligatory. But that does not matter here. All that matters is that obligatory or supererogatory, we treat competing acts here on an equal footing. Suppose we have an obligation to help a starving person in my neighborhood. If distance is not morally relevant, then we are equally obligated to help the starving far away as well as those close by. And by Singer's lights, if we have to choose, we should choose that course of action which will produce the greatest good.

But to stop here is to stop far short of the finishing line if our goal is to achieve action that instantiates such a "greatest good" general principle. Let us consider some of the ways in which a smooth and seamless connection from a greatest good prescription to action may break down. If we view decisions as being made on a case by case basis, the most obvious of these are epistemic and temporal. I may not be able to know which is the best alternative, and even if I was able to know it in principle, the time it would take to find out in practice carries costs that need to be offset against the benefits. These are the standard objections that drive a greatest good principle from being a case by case guide to a guide for formulating rules to govern classes of cases; like, in general, greater good will come from feeding the starving than freeing slaves, notwithstanding that in some cases this might not be the case.

But there is a less pragmatic and deeper traditional worry about whether a greatest good principle will in fact produce the greatest good that goes all the way back to Sidgwick (1907). We would not need a moral code in the first place if we were not prone to act immorally or at least amorally. One who advocates a moral code against this background on the grounds that it will produce the greatest good, if followed, has to offset these benefits with the costs of implementing such a policy – costs which may be high depending on the state of nature that forms its backdrop. And even that assumes success in transforming us whatever the costs. Sidgwick's concern was what the consequences will be if no full transformation is possible, if some of our natural inclinations are intractable and not fully reformable. Sidgwick's worries arise from the tension between those we are close to and strangers. By his lights, promoting the general good and the relations of "affection" we have to those close to us will come into conflict. You might think that in response, "Utilitarianism must therefore prescribe such a culture of feelings as will, so far as possible, counteract this tendency" (Sidgwick 1907, 434). But the problem is whether we can do so without transforming such feelings into "a watery kindness" (as Sidgwick points out, the phrase comes from Aristotle) when it is expressed in its universalized form. Better Sidgwick thinks, to pay the price of giving up a degree of impartiality to harness feelings for the other in their full strength even if it is only at the parochial level.

I think this picture of thinking and feeling is at best only part of the story. Of course we have special feelings for those close to us as compared to strangers. But it is not as if strangers don't evoke feelings in us as well. It is not that we lack a culture of feeling toward them. Ignoring that emotional dimension of our relations to others feeds the illusion that supplementing thinking alone is all we need to worry about when it comes to strangers. Instead, we need to wonder about how and where feelings intrude; whether in doing so they act as a help or a hindrance; and if the latter, whether they are reformable.

You are to choose between giving $100 to free two slaves or feed four people for a year. Singer whispers in your ear, "Feed the hungry, you will do more good." Set aside complex arguments about the facts. What will happen to the starving after your donation runs out? What is the probability of the slave being self-supporting after redemption? And so on. Instead, let us worry about how feelings might intrude. There are three ways. First, as an adjunct to reasoning; second, as a substitute for reasoning; and third, in the implementation of reasoning. For now let us just worry about the first two of these. Part of what drives my anecdotally based concerns about the widespread enthusiasm for slave redemption programs is the intuition that it is easier to "feel for" the enslaved than it is for the starving. And moreover, at least at first blush, these feelings bear no obvious relationship to desert.

But whether my intuition is correct or not in this particular case does not really matter because the point I want to make is a much more general one: when it comes to choosing between the objects of our moral concern it is possible to characterize at least some influences to which our feelings may be subject. That is especially the case when it comes to familiarity and distress. Under most circumstances, familiarity operates to increase feeling while distress operates to decrease it, yet neither bear a direct relationship to desert. Indeed, if anything, in the case of the latter it is even worse: the more distress inducing the case will likely be the needier. But, be that as it may, in both the cases of familiarity and distress, feelings can play a differentiating role in cases involving not just family and community versus strangers but cases involving choices between strangers alone.

If that is obvious in the case of distress, it is quite surprising in the case of familiarity, indeed even seemingly contradictory, in the sense that strangers are surely those we are unfamiliar with by definition! But to be a stranger to me is not an all or none affair. Instead they come in varying degrees of unfamiliarity. And, as we have already seen, Kunst-Wilson and Zajonc (1980) showed that familiarity of the most innocuous nature can produce a preference bias when subjects were asked to make pairwise preference judgments between random pairings of the previously exposed stimuli and novel ones (both of which were irregular octagons).

Still, someone might object as follows: we are discussing familiarity as an adjunct to moral reasoning. Surely when I ask you to choose who is most deserving, A or B, that you may prefer A over B, even without knowing it, is not necessarily to say that your moral choice will be affected. But even if mere familiarity did not have such direct effects, it can have indirect effects in virtue of its effects on empathy. For it is a truism that empathy is shaped (in part) by preferences. And if mere familiarity can drive preferences, then it will do so as well when it comes to empathy. The news here is not that, but rather that if the effects of familiarity are so fine grained that they can be exercised on choices between strangers, so too will they differentially affect our empathy for strangers. Of course feeling empathy for a stranger would be a gain to the extent that we view empathy as an unalloyed virtue. But, at least as an adjunct to moral reasoning, judgments of fairness turn out to be biased by empathy whether or not familiarity and its associated preferences bias such judgments directly or not.

Traditionally, empathy has been viewed as a necessary condition on prosocial behavior. You walk down the street and see a stranger who needs your help. Standard social psychology analyzes this sort of prima facie other directed behavior as a function of the perceived opportunity costs you face and the degree of empathy you feel (see Sears et al. 1985). It is for that reason (among others) that educational efforts to motivate and increase empathic responses are taken to serve a social good. But Batson and his colleagues (Batson et al. 1995) have shown that inducing empathy is a double edged sword when it comes to determining whom to help as opposed to whether they will be helped. Batson and his colleagues deceived student subjects into thinking they were participating in an experiment to assess the consequences of positive versus negative consequences on workers among other things. Students believed they were bring randomly assigned to either a worker or supervisor role when in fact they were all being assigned the "supervisor" role. They were then led to believe that their job was to assign the "workers" to one of two groups: those receiving positive reinforcement (a gift certificate) for successful completion of an experimental task and those receiving negative reinforcement (mild electric shock) for unsuccessful completion of the same task. Subjects were instructed that "most supervisors feel that flipping a coin is the fairest way to assign workers to the task, but it is entirely up to you" (Batson et al. 1995, 1044) and a coin was provided to them. However two thirds of the subjects were told that they would be receiving a confidential communication written by one of the people they were to assign. The content of the communication was unrelated to the assignment or task but rather talked about something sad that had recently happened in the writer's life. Of these half were told to read it but take an objective stance. The other half were instructed to read it and to try to "imagine how this student feels" (Batson et al. 1995, 1044). So in summary, this experiment examined subjects' assignments of others to positive and negative consequence groups with no communication, communication with instructions designed to elicit low empathy, and communication with instructions designed to elicit high empathy. Subjects in the low and high empathy groups were tested to establish that empathy had been elicited as intended. Results confirmed that they had. Comparing the assignments made and the methods used for the three groups: all of the "no communication" subjects used random assignments. In the "low empathy" group, 17 of the 20 subjects used random assignments and three did not, favoring the person from whom they had received the communication. However, in the high empathy group, half the subjects favored the writer. (All results reported were statistically significant.) Interestingly, nearly all participants in the study viewed random assignment as the fair way to make assignments. Moreover, those who acted partially in making the assignments viewed themselves as having acted less morally than if they had used a random assignment process.

These findings underscore how, short of universal empathy, such feelings can undercut a fair decision procedure even for subjects that are self-consciously engaged in trying to make such decisions. Choosing between outcomes in terms of the greatest good is not necessarily the same as deciding what is fair, but it too involves a deliberative process and there is no reason to think Batson's results would not apply here too. Batson's setup also nicely illustrates how easy it is to trigger empathy in a stranger and that might encourage the view that with some "juicing up" an expectation of generalized empathy is an ideal to encourage which would result in something more than watery kindness. The problem however is not just that general empathy is realizable but weak (à la Sidgwick) but rather that it is subject to catastrophic failure under circumstances when it would be most needed if it were to be relied on as an adjunct to moral reasoning. For example contrast your reactions to pictures of undernourished children, where one set is clearly underweight, gaunt, and sad looking but the second is dramatically emaciated and look to be near death. (For examples go to http://fas-philosophy.rutgers.edu/bunzl/slavery.html.)

If you are like me, images of the first kind evoke feelings of sadness in you and, if not empathy, sympathy. That empathy or sympathy may be followed by a feeling that you wished you could do something to help these sorts of children. In the case of the second set of images, you may also feel sadness, even extreme sadness and sorrow. But at the same time, you likely feel anxiety and some distress. Images like these can provoke a dramatic reaction: we may hyperventilate, physically jolt backwards, and avert our eyes. If forced to let our gaze linger, feelings of distress may increase and transform into a feeling of physical disgust. You might have had the same kind of feeling if you have ever seen pictures of a decaying corpse.

Faced with images of the severely emaciated, whether you have felt distress and stopped looking, or your feelings have reached a level of disgust immediately or after looking for a while, I suspect you have felt less empathy or sympathy than you have when confronted with images of the "merely" undernourished. And you have done so even as you may have thought that the former were more deserving of your sympathy. If I am right, you also did not have a feeling that you wished you could do something to help the severely emaciated. Your level of distress and, if it happened, your disgust reaction trumped that. If you have had such a reaction, note how it distanced you from your connection to the images. You may have caught yourself thinking that you did not want to think about these images, or thinking that I had a lot of nerve to have you think about them, or having a myriad of other self-focused thoughts. Distress and disgust functioned to break your connection with the needs of others, replacing it with your own needs.

But if all of this makes empathy seem a dubious ally for moral reasoning, let alone, a substitute for such reasoning, it isn't as though we can ignore its vagaries, if only as a source of interference with reasoning. And if the traditional worry (going back to Sidgwick) is that such feelings intrude only close to home, here we can see that its reach is much further than we might think. Now one way to react to this is to argue for a traditional rationalist line that we should eschew a Humean conception of morality as a slave of the passions and boldly assert the preeminence of reasoning over feeling where the two conflict. Of course that assumes we can discipline our feelings at least when it comes to reason. But we better not hope for too much success if we are interested in the probability of the conclusions of such reasoning being implemented, at least on the standard model of altruism outlined above. For on that model empathy is a necessary condition for us to *act* altruistically as opposed to merely reasoning about it. That creates a new opportunity for empathy to undermine the outcome of reasoning. Not directly, but instead by the selectivity by which we are responsive to its prescriptions.

III

Let us assume empathy is a necessary condition for altruistic action without worrying too much about the alleged distinction between empathy and sympathy. How deferential should we be to empathy's dictates? Should the moral prescriptivist hold his or her ground in the face of empathically driven insurrection or settle for the next best thing? Even if we assume the model of altruism we are relying on is correct, we can't begin to answer these questions without a sense of how reformable empathy is. I think there is reason to be pessimistic about the prospects for reforming empathy given the nature of the kinds of factors that we have seen that can affect it. If the truly needy are unfamiliar (in the way discussed above) and disgust provoking in their very neediness, then empathy comes under attack from both directions and its chances of prevailing in provoking action will suffer. But even bracketing these external disruptions to our empathic capacities, empathy itself contains the seeds of its own destruction. If altruistic action has empathy as a necessary condition, it is not just any empathy but empathy for or with the distress of the other. I experience what you feel. My feelings match yours. But for me to act altruistically I have to respond to those feelings on your behalf. On Martin Hoffman's view (2000) my capacity to act in response to such empathic distress eventually gets undermined by a self-directed reaction when the empathic distress reaches a certain level. Notice that this can function parallel to the feelings of distress or disgust which the needy (like those in the images of the severely emaciated) elicit from most of us directly. Those feelings block empathy. Here empathy draws us to another, but if we get too drawn to them, and their condition is too extreme, our other directed response tips a focus on ourselves and withdrawal from the other. Thus, the greater your distress, *and* the greater my empathic response to it, the greater the chance that I will act on my behalf not yours.

Unlike disgust reactions, there is no reason to think that the tipping point from empathetic concern for the other to concern for oneself is not quite varied and subject to many determinants. Be that as it may, with the exception of Mother Theresa, for most of us, moral prescription and empathic responses will part ways at some point. Here a moral prescriptivist would do well to think like a preacher. A preacher, at least one interested in producing good, ought to worry about whether his or her prescriptions fall within the realm of psychological possibility for his or her charges. That is not to say that there are not religious orders that render us always wanting. But a religion that makes the task of moral improvement hopelessly unattainable runs the risk of provoking the embrace of sin. After all, if we are doomed to be sinners, why not at least have fun along the way?

Of course the danger, for a preacher or prescriptivist, or the temptation (depending on your view) is to misjudge psychological possibility. If aiming too high makes us all and only sinners, aiming too low makes it too easy for us to be saints. Yet should we be concerned about possibility or would we do better to rather think about probability? Not, obviously, just the probability of the course of action I am likeliest to engage in, or even the product of the probability of such an action and the good it produces. Nor just the fact that people are more likely to free a slave than feed a starving person, even if the latter would do more good without probabilities in the picture. Instead, what we should worry about is the probability of people being capable of being induced to do the latter over the former. That is to say, whether people's probabilities of action to bring about the outcome that will produce the greatest good can be increased. If you are interested in producing the greatest good, this choice turns out to be quite simple, but only if you have all the facts in hand. You or Singer will want to know the effect of your prescription on increasing the probability of my engaging in one action rather than another, along with the good resulting from those actions. Perhaps in the end advocating the second best will produce the greatest good.

But this is surely madness. We have reduced the role of a moral prescriptivist to a calculational opportunist. Instead of advocating what Singer believes to be the best course of action, we want him to advocate the best course of action in the light of what we are likely to do in the light of his advice, even if it is second best. It is one thing for the prescriptivist not to advocate a courses of action that I am not capable of following but another to specify a course of action for me based on the belief that specifying so will produce the best outcome in the long run. But why? The answer is that the second loses sight of my agency and treats me merely as an extension of your or Singer's agency. But that said, prescribing with an eye on the range of my psychological capacities makes sense because those capacities act as limits on my agency. If that sounds like a plea for ought to respect the limits of can, before we even worry about such arguments, we should dwell on the idea of capacity here.

It is one thing to say there is no point in advocating a course of action that would require me to be in two places at once because that is physically impossible. Is it another to say there is no point in advocating a course of action that would require me to sacrifice my child to save a strange child? Or at least to toss a coin? Psychological possibility comes here in two grades; my capacities as they are now, given who I am and the range of capacities I could have. Given who I am now, I would find it impossible (psychologically) to do many things that stronger people might be able to do, and that I might be able to do if I were to become a stronger person. Assume that second set of possibility has limits that are perhaps species specific. Do these even come close to the sense of physical impossibility that grounds moral arguments that to hold one responsible for a failure to act it had to actually be possible for that person to act? But notice that even if they do, nothing discussed here so far (including disgust reactions) comes close to this. For even in the case of disgust, we have to allow that some people are capable of overriding their disgust reactions (most notably, Catherine of Siena, see Miller 1997).

So, am I driven back to calculational opportunism as the only reason for advocating the second best? Is that the only reason to prescribe freeing slaves over feeding the hungry? To do so is to accept the picture as one that sees the matter as respecting our agency or behaving opportunistically. But that seems to me to set up a false dichotomy. The tradition of moral reasoning begun by Piaget, and developed by Kohlberg, treats (nearly) all of us as capable of moral reasoning with varying degrees of sophistication. But in addition to its disinterest in the consequences of such reasoning for moral action, it treats moral reasoning as a process that is isolated from the effects of other psychological process. The whole enterprise is a purely internal matter. We tend to treat agency as an artifact of this reasoning process and emotion as an interference – what Elster terms, "sand in the machinery of action" (Elster 1998, 284). That picture feeds the idea that with discipline we can filter out that interference. But what if that is an illusion? Some (like Peter Railton) would turn this into a virtue, arguing that it is a hopeless enterprise for us to hope to follow the dictates of consequentialism directly and explicitly. Instead, we should aim to harness our psychology through collective action that creates a culture of values that serves the goal of maximizing the good:

> A sophisticated act-consequentialist should realize that certain goods are reliably attainable – or attainable at all – only if people have well developed characters; that the human psyche is capable of only so much self-regulation and refinement; and that human perception and reasoning are liable to a host of biases and errors. Therefore individuals may be more likely to act rightly if they possess certain enduring motivational patterns, character traits, or prima facie commitments to rules in addition to whatever commitment they have to act for the best.
>
> (Railton 1984, 158)

But, whether you view agency expansively as encompassing these emotions, or narrowly but as always operating with them in the background, understanding their vagaries, plasticity, and educability is unavoidable if we are interested in understanding our capacity for moral action. Only then can we know if the next best thing is the most we can reasonably be expected to do.

IV

I have framed these considerations within the context of the competing claims of the enslaved and the starving. But do they have an echo in the competing claims between us in the here and now and those (should they exist) in the future? At first blush it would seem that the situation is very different. For as we saw earlier, our relationship to those in the future, even those genetically related to us, quickly fades into abstraction. What begins as a rich textured emotional fantasy about my grandchildren-to-be thins as I think of their children – and then theirs – and so on. Just as my inclination to privilege my grandchildren to be over strangers loses its force as I think of doing the same for their children and so on. And if I think of your grandchildren and their children and so on, indifference rears its heads much further up the chain of being. But *is* that so different from how we feel about the starving? You are told a million people are starving in southern Somalia. My reaction to them is (I hate to say) one of indifference as well.

Two things drive that indifference, their number and their distance. Which is why fundraisers for the starving traffic in personal stories, the most effective of which are told by "survivors" who visit us (literally or figuratively) in our living rooms. But absent the discovery of time travel, that option is not open to fundraisers for the future stuck where they are and vast in number.

Moreover, when it comes to doing the second best, as in saving a slave rather than the starving, even if doing the latter would do more good, a deeper difference arises. For in choosing the present over the future, except for the considerations raised about the poor in Chapter Four, we are not here confronted with two good acts. Instead we think of favoring us now over them then as a selfish choice. In this case empathy does not align with a moral choice, even if it is not the best moral choice, but rather aligns with a choice that is not a moral choice at all. And if that is the case, and we favor moral choice, there is no second best choice. Hence our only option will be to work to strip the force of empathy from our decision making since it stands in an antagonistic relationship to the actions we ought to take. At least it is, unless you think we could somehow "juice" up our feelings toward those in the future and imbue our relationship with empathy for them. That seems to set the stage for movie makers to make their entry bearing dystopian visions of the future, or for fiction writers to pen heartfelt pleas in the voice of those who will follow us. But you only need to watch the movie *Waterworld* or read Clara Hume's (2012) *Back to the Garden* to understand the psychological limitations of this as a strategy. We may be moved emotionally, but most of us are not moved when it comes to action. In the final analysis, the medium of fiction limits the effects of its messages in this regard. So, again short of actual time travel, the idea that we can infuse our relationship with those in the future with feeling does not look promising. In the end, a "watery" kindness is all we should expect and hope for.

Is the only option then to decathect our feelings toward those in the here and now as a way to level the psychological playing field? Not necessarily. There is another option that we might embrace.

Elke Weber and her colleagues at the Center for Research on Environmental Decisions at Columbia have pioneered studies that examine what and when we are moved when it comes to warnings about the pending danger of climate change. In her work, Weber (Weber 2010, 2011) generalizes to assert that personal experience of the perceived effects of climate change is a more important driver of risk management than time delayed, abstract, statistical data. As a consequence, the closer to home the content of a message is, the more likely we are to be responsive to it. Weber's research echoes the idea of second best moral choices that may be more likely to elicit moral action than the best moral choice. But here the moral dimension does not play a role. All that may seem to be in play is the idea that the best way to move someone to action may not be on the basis of the most pressing scientific reason. Forget about future generations and focus on the here and now for that will be more likely to move you as it were even if the idea that it does sticks in our craw. For future generations climate change will mean the end of life as we have known it, and, as I have been arguing, perhaps the end of life altogether. But for most of us, in the here and now, climate change will merely mean a limited change in our life as we know it: hotter summers, wetter winters, more power outages, and so on. Life will carry on as it was with minor adjustments. If the former cries out as the primary reason we ought to act, it is the latter that seems more likely to prompt us to act.

But the problem with this picture is that it too assumes that our interests and those of the other are in harmony when it comes to avoiding climate change. Their need to avoid it may be more pressing than ours, but nonetheless we would do better to cast the reasons in our terms not theirs. But I have argued that our situation is not this. Ours is one in which our interests and the interests of those who will follow us are (by and large) in conflict. ("By and large" because of course there is no denying that some in the here and now will be dramatically affected by climate change in the short run as well, even if they are comparatively small in number. But like future generations, they too will stand apart from the felt experience of most of us.) Nonetheless, despite this difference there is a lesson we can draw related to Weber's research. To the extent we can represent the interests of those far off from us, be it in space or time, in the lived reality of our world, we improve the chances of our being responsive to their needs and being so in a thick textured way. But, as we just noted, that is easier said than done, especially when it comes to future generations. Given that, what hope is there for finding a voice in the present for those in the future?

But if not that, then what? To be concerned about the welfare of those in the future does not require us to express that concern in those terms. But then in which terms, given the antagonism of their interests and ours? Of course the obvious alternative here is to ask if we can diminish that antagonism by bringing our interests more into harmony with theirs. But much of the foregoing has been devoted to arguing that such a strategy is not as self-evident as might seem at first, both philosophically and psychologically. Still if those arguments stand, can we at least appear to bring our interests and theirs into harmony? That seems like a call to sacrifice truth for the greater good. For the truth of the matter is that the short-term features of climate change (in our lifetime and those of our children) are not likely to be drastic enough to upend a calculus of self-interest that points in the other direction. Instead, a much more promising and honest alterative would seem to me to take seriously the arguments for transcendentalism examined earlier but apply them not to all of nature, which failed, but just to our own kind.

Transcendentalism on this variant invites the idea that we efface the boundaries of the self with each other across time in an attempt to expand the reach of our self-interest. The end goal of the resulting fusion is the creation of a unitary subject that reaches into the future, albeit restricted to our own species. As I suggested earlier, perhaps with transcendentalism in hand we would be less prone to act on our narrow short-term interests. But I worried that this put the cart before the horse. With our short-term interests fixed, those are the interests which will govern our attitude toward the other; be it nature in general, or future generations of our own species. And that attitude might include embracing the idea of there being no future generations.

But are those interests necessarily fixed? If we take the idea of fusion seriously, we should be open to the possibility of a dialectic which allows our expanded notion of who we are to upend those very interests, and come to care about future generations not out of altruism but because they are us.

"Easy for you to say!" yells Nguyen Thi Luu, one of a number of single mothers living in the village of Loi, Vietnam. Some were widowed during the war with the United States, others never married. All have had planned pregnancies successfully breaking a cultural taboo against doing so. But this is not simply a desire for the pleasure of raising children. "I was afraid to die alone," Julie Cohn quotes Ms. Luu as saying. "I wanted someone to lean on in my old age" (Cohn 2013). To answer her complaint, transcendentalism has to start much closer to home. It has to embrace an expanded notion of who we are if we are to come to care not just about future generations but also about our own generation, in the here and now too – not out of altruism, but because they are us as well.

References

Batson, Daniel, Tricia Klein, Lori Highberger and Laura Shaw 1995. "Immorality From Empathy-Induced Altruism: When Compassion and Justice Conflict," *Journal of Personality and Social Psychology*, 68 (6): 1042–1054.

Cohn, Julie 2013. "A Tiny Village Where Women Chose to Be Single Mothers," *New York Times*, February 17, www.nytimes.com/2013/02/15/world/asia/in-vietnam-some-chose-to-be-single-mothers.html?_r=0, accessed February 17th, 2013.

Elster, Jon 1998. *Alchemies of the Mind*, Cambridge: Cambridge University Press.

Hoffman, Martin 2000. *Empathy and Moral Development*, Cambridge: Cambridge University Press.

Hume, Clara 2012. *Back to the Garden*, Coquitlam, BC: Moon Willow Press.

Karlan, Dean and Alan Krueger 2007. "Some Simple Analytics of Slave Redemption," in K. Anthony Appiah and Martin Bunzl (eds.), *Buying Freedom: The Ethics and Economics of Slave Redemption*, Princeton, NJ: Princeton University Press, 9–19.

Kunst-Wilson, William and Robert Zajonc 1980. "Affective Discrimination of Stimuli That Cannot Be Recognized," *Science* 207: 557–558.

Miller, William 1997. *The Anatomy of Disgust*, Cambridge: Harvard University Press.

Railton, Peter 1984. "Alienation, Consequentialism, and the Demands of Morality," *Philosophy and Public Affairs*, 13(2): 134–171.

Sears, David, Jonathan Freedman and Lelita Peplau 1985. *Social Psychology*, New York: Prentice Hall.

Sidgwick, Henry 1907. *The Methods of Ethics*, London: Macmillan.

Weber, Elke 2010. "What Shapes Perceptions of Climate Change?," *Wiley Interdisciplinary Reviews: Climate Change*, 1(3): 332–342.

Weber, Elke 2011. "Climate Change Hits Home," *Nature Climate Change*, 1: 25–26.

INDEX

Note: Page numbers with *f* indicate figures; those with *t* indicate tables.

CPSIA information can be obtained
at www.ICGtesting.com
Printed in the USA
FFOW02n1343260116
20848FF